芯片验证调试手册

验证疑难点工作锦囊

刘 斌 著

电子工业出版社
Publishing House of Electronics Industry
北京·BEIJING

内 容 简 介

资深芯片验证专家刘斌（路桑）围绕目前芯片功能验证的主流方法——动态仿真面临的日常问题展开分析和讨论。根据验证工程师在仿真工作中容易遇到的技术疑难点，本书内容在逻辑上分为 SystemVerilog 疑难点、UVM 疑难点和 Testbench 疑难点三部分。作者精心收集了上百个问题，给出翔实的参考用例，指导读者解决实际问题。在这本实践性很强的书中，作者期望能够将作者与诸多工程师基于常见问题的交流进行总结，以易读易用的组织结构呈现给读者，目的是帮助芯片验证工程师更有效地处理技术疑难点，加快芯片验证的调试过程。

本书面向在岗的芯片验证工程师，可作为日常桌边工作手册翻阅，也可用于工作之余查漏补缺以提高自身技术能力。

未经许可，不得以任何方式复制或抄袭本书之部分或全部内容。
版权所有，侵权必究。

图书在版编目 (CIP) 数据

芯片验证调试手册：验证疑难点工作锦囊 / 刘斌著. —北京：电子工业出版社，2023.1
ISBN 978-7-121-44845-4

Ⅰ. ①芯⋯ Ⅱ. ①刘⋯ Ⅲ. ①芯片－验证－手册 Ⅳ. ①TN43-62

中国国家版本馆 CIP 数据核字（2023）第 005446 号

责任编辑：窦　昊
印　　刷：河北虎彩印刷有限公司
装　　订：河北虎彩印刷有限公司
出版发行：电子工业出版社
　　　　　北京市海淀区万寿路 173 信箱　邮编：100036
开　　本：787×1092　1/16　印张：15　字数：384 千字
版　　次：2023 年 1 月第 1 版
印　　次：2025 年 5 月第 5 次印刷
定　　价：69.00 元

凡所购买电子工业出版社图书有缺损问题，请向购买书店调换。若书店售缺，请与本社发行部联系，联系及邮购电话：(010) 88254888，88258888。
质量投诉请发邮件至 zlts@phei.com.cn，盗版侵权举报请发邮件至 dbqq@phei.com.cn。
本书咨询联系方式：(010) 88254466，douhao@phei.com.cn。

自　　序

距离着手准备出版《芯片验证漫游指南》已经有将近五年之久，回想当时在校对书稿时有一种孕育已出的感觉。而今再次准备新书时，当时的感觉又一次涌上心头，一样的喜悦是出版的内容都将会让读者们受益，而不一样的时间点让我后来改变了出版计划，放下另一本关于系统验证思想的书稿计划，将本书先行出版。

回想2017年时，芯片行业和验证领域都还未有如今之繁荣景象，无论是企业、资本，还是人力、产品，今日都是一派人潮涌动。在这一背景下，越来越多的芯片验证工程师进入验证领域，而从行业的人力经验构成来看，我们似乎还需要更多的时间和耐心让后续加入的这一批批新人快速成长，担以重任。

相比于系统验证思想的大局观，刚入局不久的验证工程师在面临项目进展和技术学习的双重压力时，更多地需要解决眼下的技术疑难点，才能稍有时间去补充技术知识。早前芯片行业人才储备的不足，使得在最近几年有大批年轻验证工程师涌入，在此情况下，芯片行业传统的老带新、师傅带徒弟的隐性行规比较难以执行下去。在人才市场重新组合之后，资深的验证工程师进入了初创公司，迫于项目压力，有时也难以顾及系统性的新人培养。

芯片行业以往持续进行的这种经验传授，由于人力结构的骤然变化目前已经受到挑战。我在当时准备《芯片验证漫游指南》一书时，已将传道授业解惑贯穿于中。现如今与上千位在路科学习的验证工程师沟通时，焦点也多以工作具体问题为主。基于此，其觉得应先出版一本能够解决年轻验证工程师燃眉之急的疑难点手册。主意定下来之后，我便从2021年春开始陆续收集、总结这些验证知识疑难点，先以每日验证知识小卡片的问答形式传播于路科学习交流群，而后再扩展至路科验证订阅号在节气时推送的验证知识小卡片，并进行总结。

这样的问答形式在当时受到了学员们和朋友们的肯定，也鼓励我在2022年春季继续将以前的和后续收集的问题进一步完善，并配以更加完整的代码示例，以期能够达到出版的标准。我深知芯片行业的技术交流还受限于诸多方面，验证工程师们的交流也可以更加活跃。本书将在目前这个阶段，帮助验证工程师们查漏补缺，实实在在地解决他们工作中的疑难点。

需要承认的是，目前国内在芯片验证领域，处于主流的、从业人数最多的仍然是功能仿真验证技术，本书也将围绕SystemVerilog、UVM和Testbench这三个大类将疑难点分别归类，以便合理地将问题组织起来，方便查阅。

同时具备芯片验证领域的企业咨询、职业教育和高校授课的经验，让我能够有更广的视角、更敏锐的嗅觉，理解企业、求职候选人和验证工程师的需求，也因此在目前这个阶段，选择出版这本书，以期帮助验证工程师们解决日常问题。

由此看来，路科验证和我希望能够在这个芯片行业的变局中，在这个芯片人才的迁移浪潮中，**扮演蜂巢的角色**，为行业培育合格的验证工程师，陪伴他们度过验证新人的成长阶段，开展多种形式的验证技术和职业交流，持续帮助他们提升综合能力。

时刻躬行，不时反问自己，是否在日进一寸。唯有自身日进，才能持续输出，在芯片验证领域传播知识，帮助更多的验证工程师。

2022 年 7 月 11 日
作于 西安 软件新城

目　　录

第1章　什么是疑难点？ 1
第2章　SystemVerilog 疑难点集合 2
2.1　数据使用 2
　　2.1.1　组合型数组和非组合型数组怎么区分？ 2
　　2.1.2　组合型数组和非组合型数组如何做赋值？ 2
　　2.1.3　在使用 enum 或 struct 时添加 typedef 与否的差别是什么？ 4
　　2.1.4　什么是静态变量和动态变量？ 6
　　2.1.5　struct 和 struct packed 区别在哪里？ 6
　　2.1.6　关联数组的散列存储表示什么？ 7
　　2.1.7　如何将队列插入到另外一个队列中？ 8
　　2.1.8　队列在赋值时使用操作符{ }，那么它属于组合型吗？ 8
　　2.1.9　数组的选取可以用两个变量作为索引边界吗？ 9
　　2.1.10　parameter、localparam 和 const 有什么联系和差别？ 10
　　2.1.11　多维数组的声明和使用，哪种方式更合适呢？ 11
2.2　操作符使用 14
　　2.2.1　{ }操作符的使用场景有哪些？ 14
　　2.2.2　条件赋值符?:和条件语句 if-else 的执行结果一致吗？ 15
　　2.2.3　if 和 iff 的应用场景分别有哪些？ 16
　　2.2.4　使用 foreach 在轮循数组时按照什么顺序呢？ 17
　　2.2.5　运算符的优先级是否有必要记忆呢？ 18
　　2.2.6　assign 连续赋值可以赋值给 logic（var）类型吗？ 19
　　2.2.7　:: 和 . 这两个符号使用起来有哪些区别？ 20
2.3　模块、接口与方法 22
　　2.3.1　module 中的方法在声明时是否要添加 automatic？ 22
　　2.3.2　interface 在何处需要使用 virtual 来声明呢？ 24
　　2.3.3　initial 和 always 的执行顺序是否与代码位置有关？ 25
　　2.3.4　interface 的 modport 和 clocking block 如何使用？ 26
　　2.3.5　module 和 interface 之间可以相互例化吗？ 27
　　2.3.6　方法的参数默认方向该如何辨别？ 28
　　2.3.7　return 的使用场景有哪些？ 29
　　2.3.8　task 与 function 的联系和差别在哪里？ 30
　　2.3.9　方法的参数默认值该如何使用？ 31
　　2.3.10　方法中参数方向 inout 和 ref 有什么差别？ 32

2.3.11 module 和 interface 中的变量声明必须放置在头部位置吗? ………… 33
2.3.12 如何例化和传递多个相同类型的接口? ………… 35
2.3.13 使用 while 和 forever 语句时需要注意什么? ………… 36
2.3.14 系统函数和内建方法有什么区别? ………… 37
2.3.15 接口和模块的联系和差别是什么? ………… 38
2.3.16 program 和 module 的联系和差别是什么? ………… 38
2.3.17 多个线程在仿真调度中是如何在不同的域之间执行和切换的? ………… 39
2.3.18 时钟块在使用时需要注意哪些地方? ………… 39
2.3.19 如何连接和驱动双向端口信号? ………… 40

2.4 类的使用 ………… 43
2.4.1 类的成员变量在声明时或在 new 函数中初始化有何区别? ………… 43
2.4.2 new 函数与其他函数有哪些不同? ………… 44
2.4.3 关键词 new 的使用场景有哪些? ………… 45
2.4.4 对象占用的空间什么时候会被回收? ………… 46
2.4.5 this 的使用场景有哪些? ………… 46
2.4.6 对象拷贝的种类如何划分? ………… 47
2.4.7 this 和 super 有什么联系和差别? ………… 49
2.4.8 子类成员如何覆盖父类成员? ………… 50
2.4.9 子类句柄和父类句柄的类型如何转换? ………… 52
2.4.10 为什么父类句柄经常需要转换为子类句柄? ………… 53
2.4.11 虚方法声明和不声明的区别是什么? ………… 54
2.4.12 虚方法的描述符 virtual 应该在哪里声明? ………… 55
2.4.13 句柄一旦不指向对象,该对象就被回收了吗? ………… 58
2.4.14 什么时候该使用 local 和 protected? ………… 58
2.4.15 class 和 virtual class 的区别是什么? ………… 58
2.4.16 virtual 修饰符在哪些场景中会用到? ………… 60
2.4.17 子类能够使用与父类相同名称但不同参数的方法吗? ………… 61
2.4.18 使用参数类或者接口时要注意哪些? ………… 62

2.5 随机约束使用 ………… 64
2.5.1 rand 描述符可用于哪些变量类型? ………… 64
2.5.2 数组使用 rand 声明会发生什么? ………… 65
2.5.3 句柄使用 rand 声明会发生什么? ………… 66
2.5.4 rand 和 randc 的区别在哪里? ………… 67
2.5.5 内嵌约束中的 local:: 表示什么? ………… 68
2.5.6 是否可以利用动态数组对变量值的范围进行约束? ………… 70
2.5.7 多个软约束在随机化时有冲突是否可以解决? ………… 71
2.5.8 结构体是否可对其成员使用 rand 描述符? ………… 72
2.5.9 如何随机化对象的多个成员且使每次数据不重复? ………… 73
2.5.10 子类会继承还是覆盖父类的约束? ………… 74

2.6 覆盖率应用

- 2.6.1 covergroup 的采样事件如何指定？……77
- 2.6.2 covergroup 如何对变量进行采样？……78
- 2.6.3 是否可对 covergroup 中的不同 coverpoint 指定采样条件？……78
- 2.6.4 covergroup 在哪里定义和例化更合适？……80
- 2.6.5 如果 covergroup 中的 bins 没有被采样，可能有哪些原因？……81
- 2.6.6 如何减少不关心的 cross bins 采样数据？……83

2.7 线程应用

- 2.7.1 semaphore 使用时需要初始化吗？……87
- 2.7.2 mailbox 使用时需要例化吗？……88
- 2.7.3 fork-join_none 开辟的线程在外部任务退出后也会结束吗？……90
- 2.7.4 父线程和子线程之间的执行关系是什么？……91
- 2.7.5 disable fork 和 disable statement 有什么差别？……93
- 2.7.6 嵌套的 fork 有没有可能被 disable fork 误伤呢？……94
- 2.7.7 使用 for 配合 fork-join_none 触发多个线程时需要注意什么？……96

2.8 断言应用

- 2.8.1 SV 语言如何控制断言的开关？……98
- 2.8.2 仿真器如何控制断言的开关？……99
- 2.8.3 断言在哪里定义和例化更为合适？……101
- 2.8.4 如何更好地让接口中的断言实现复用性？……102

第3章 UVM 疑难点集合 ……104

3.1 UVM 机制 ……104

- 3.1.1 是否所有的 UVM 对象都应该用工厂创建呢？……104
- 3.1.2 工厂创建 uvm_object 是否需要为其指定 parent？……106
- 3.1.3 为什么建议配置放在对象创建之前？……108
- 3.1.4 UVM 环境中进入 new() 和 build_phase() 有什么区别？……108
- 3.1.5 在创建组件时采用 new() 有什么影响？……110
- 3.1.6 UVM 配置类的参数修改应该发生在什么时间？……111
- 3.1.7 UVM 的消息严重等级是否可以屏蔽或修改触发动作？……112
- 3.1.8 UVM 的消息严重等级是否可以修改？……112
- 3.1.9 通过 uvm_config_db 可以完成哪些数据类型的配置？……115
- 3.1.10 使用 uvm_config_db 时传递的参数类型是否需要保持一致？……116
- 3.1.11 如何设置 timeout 时间防止仿真超时？……119
- 3.1.12 set_drain_time() 的作用是什么？……119
- 3.1.13 组件的 phase 方法中继承父类的 phase 方法是在做什么？……121
- 3.1.14 如何控制 UVM 最后打印的消息格式？……122
- 3.1.15 配置对象的层次为什么应与验证环境的层次相同？……123
- 3.1.16 uvm_config_db 和 uvm_resource_db 在使用时有什么区别？……123
- 3.1.17 在继承 UVM 某些参数类时是否需要指定参数类型？……124

- 3.1.18 UVM 中的注册类有重名时会发生什么？ ························125
- 3.1.19 为什么在 build_phase 中访问更低层次的组件会失败呢？ ········128
- 3.1.20 在 UVM 中是否可跨层次调用某些组件的方法？ ················130
- 3.1.21 在使用 uvm_config_db 时要注意什么？ ························133

3.2 通信与同步 ··136
- 3.2.1 event 和 uvm_event 的联系和区别是什么？ ·······················136
- 3.2.2 sequencer 和 driver 的类型参数 REQ/RSP 需要保持一致吗？ ······136
- 3.2.3 TLM FIFO 的方法是否可以直接调用？ ·····························137
- 3.2.4 为什么 uvm_object 类型不能例化 TLM 端口？ ····················139
- 3.2.5 UVM 回调函数类的使用特点有哪些？ ·····························140
- 3.2.6 uvm_event 应从哪里例化和获取？ ·································140
- 3.2.7 TLM 端口为什么没有注册过呢？ ···································142

3.3 测试序列 ··144
- 3.3.1 m_sequencer 和 p_sequencer 有什么区别？ ·······················144
- 3.3.2 为什么不建议在 sequence 中使用 pre_body()和 post_body()？ ······146
- 3.3.3 sequence 如何通过 uvm_config_db 获得配置的变量？ ···············147
- 3.3.4 start()和`uvm_do_on()有何区别？ ·································149
- 3.3.5 uvm_sequence_library 的作用是什么？ ······························149
- 3.3.6 配置 default_sequence 和调用 sequence::start()是否可同时进行？ ····152
- 3.3.7 一些 sequence 调用 raise_objection()的目的是什么？ ···············152
- 3.3.8 每一个 sequence 都需要调用 raise_objection()吗？ ···············154
- 3.3.9 set_automatic_phase_objection()使用起来方便吗？ ················155
- 3.3.10 如何终止一个正在执行的 sequence？ ·····························157
- 3.3.11 发送 sequence 和 sequence item 的优先级问题是什么？ ············159
- 3.3.12 为什么 sequence 通过 get_response()可以得到正确的 response？ ·····160
- 3.3.13 通过 uvm_config_db::set()或 start()挂载 sequence 有哪些联系和差别？ ···161
- 3.3.14 通过 uvm_config_db 挂载 default sequence 需要注意什么？ ········164
- 3.3.15 为什么不建议使用 default_sequence 挂载顶层序列呢？ ············166
- 3.3.16 uvm_sequence::start()挂载的 sequencer 什么情况下需要指定？ ·····167
- 3.3.17 virtual sequence 需要获得某些信号和状态应该如何实现？ ·········168
- 3.3.18 怎么让 sequence 感知 coverage 的增长并及时停止呢？ ············168

3.4 寄存器模型 ··173
- 3.4.1 寄存器模型验证常见的测试点有哪些？ ······························173
- 3.4.2 使用 set_auto_predict()和 predictor 更新寄存器有什么区别？ ·······173
- 3.4.3 如何对寄存器的某些域进行读/写操作？ ·····························174
- 3.4.4 uvm_reg_filed::configure()中的参数 volatile 的作用是什么？ ·······175
- 3.4.5 uvm_reg 的回调函数 {pre, post}_{write ,read} 的用途是什么？ ········177
- 3.4.6 与 uvm_reg_cbs 相关的回调函数的用处是什么？ ····················178
- 3.4.7 adapter 中的 provides_responses 属性的作用是什么？ ···············179

        3.4.8　多个 uvm_reg_block 和 uvm_reg_map 的关系如何影响对寄存器的访问？·················180
        3.4.9　如果并行利用 RGM 对寄存器做读/写可能出现什么问题？·································181
        3.4.10　寄存器模型结构是否支持多个 top RGM？···181
        3.4.11　uvm_reg_map 的数据位宽如果与总线不同需要做什么处理？·······························182
        3.4.12　uvm_reg_map 的数据位宽如果与子一级不同需要做什么处理？···························184
        3.4.13　寄存器模型的镜像值和期望值什么情况下相等或不相等？····································186
        3.4.14　uvm_reg 的读/写动作在发起后没有结束的原因可能是什么？·······························187

第 4 章　Testbench 疑难点集合···189

4.1　编译与导入··189
        4.1.1　package 中可以定义什么类型？···189
        4.1.2　library 和 package 怎么区分？··190
        4.1.3　文件中出现 typedef class X 是什么意思？··191
        4.1.4　`include 和 import 的差别在哪里？··192
        4.1.5　`include 应该在哪里使用？···193
        4.1.6　应该怎么理解域（scope）呢？···194
        4.1.7　在系统验证阶段如何避免反复编译以节省时间？··195
        4.1.8　如何解决和避免类重名或模块重名的问题？··196

4.2　验证组件实现···197
        4.2.1　监测器采样数据需要注意哪些？···197
        4.2.2　模块中的信号可以强制赋值和监测吗？··197
        4.2.3　如何对设计层次中的某个实例进行侵入式赋值？··199
        4.2.4　如何在仿真结束时打印一些测试总结信息？··201
        4.2.5　为什么有时无法在 sequence 或 test 中使用 force 语句？···203
        4.2.6　为什么寄存器模型机构应与验证环境层次保持一致？··205
        4.2.7　为什么建议执行任务时各组件统一使用 run_phase 或 main_phase？·······················205
        4.2.8　如何更新 driver 的驱动行为？···209
        4.2.9　force 和 $hdl_xmr_force、uvm_hdl_force 等命令有什么差别？·································210

4.3　测试平台控制···212
        4.3.1　如何将覆盖率数据信息与测试用例关联？··212
        4.3.2　系统验证如何实现 C 用例和 UVM 用例之间的互动？···212
        4.3.3　系统验证时测试用例有误，是否可以避免重新仿真而只做局部修改？·················213
        4.3.4　如何在仿真过程中更好地控制验证环境的结构和行为？··214
        4.3.5　不同目标具备不同 timescale 是否合适？··215
        4.3.6　能否对其他组件执行的 raise objection 强行操作 drop objection？···························217
        4.3.7　如何在回归测试中减少冗余的测试用例？··219
        4.3.8　验证环境遇到 reset 时如何协调各验证组件？··220
        4.3.9　仿真出现错误信息时如何让仿真停止？··224
        4.3.10　Verilog 如何实现在相同结构中采用不同设计模块？··225
        4.3.11　仿真器如何实现在相同的结构中采用不同的设计模块？······································227

第 1 章

什么是疑难点？

相比于 SystemVerilog 的语法和 UVM 的框架规范，在利用这两者搭建测试平台的过程中，仍然会出现一些对于芯片验证工程师而言"稀奇古怪"的问题。如果是 SystemVerilog 的语法问题，那么验证工具在编译时可以立刻告诉我们这一点。如果是 UVM 的框架规范问题，我们也可以在搭建测试平台之前，根据公司已有的框架建议去遵循，以便可以与其他验证框架保持一致。

而验证技术的疑难点，还无法用语法和框架规范去覆盖。验证技术的疑难点我们在本书中将其划分为三大类，分别是 SystemVerilog 疑难点、UVM 疑难点和 Testbench 疑难点。这些疑难点大多来自日常工作中遇到的疑问和困难。工作中的疑问，会让这些问题一直藏在代码中，直到某一天暴露出来，而验证工程师才发现，他之前对该疑问的理解不够全面，甚至推翻了他对这一问题的理解；工作中的阻碍，又会常常让验证工程师受困于某个技术问题，在不能找到合适的解决方法之前，他可能受迫于项目压力，找了其他临时的替代办法绕过了这个问题，而这个问题在他没有解决之前，将会一直成为他的一个技术盲点。

如果我们既可以从理论、语法着手来解释一些疑问，也可以从一些实际代码用例来解决一些技术困难，那么这将为数以万计的验证工程师带来实际帮助。本书中的疑难点，一方面来自作者与验证工程师们交流问题时的总结，另一方面来自作者从事验证技术咨询时，帮助客户解决的问题。

基于与路科紧密联系的上千位工程师每日的工作问题交流，作者将近 170 个问题和近 100 个参考用例以合适的组织结构展现给读者。可以预见的是，在未来的几年中，作者还将继续收集代表性的疑难点，并持续更新本书的内容。本书的初衷是帮助验证工程师有效地处理好日常技术疑难点，尤其是当他身边缺少一位可以商量技术问题的伙伴时。

读者在阅读本书时，既可以利用休息的片段，找到自己感兴趣的疑难点翻阅，也可以在遇到某个问题时，到书里试着找到解决问题的线索。读者也可以用完整的时间从前到后翻阅这本书。如果读者的问题暂时没有能得到解决，可以将问题发送到邮箱 bin.liu@rockeric.com。这些好的问题，也将持续被收录到书中，用来帮助验证工程师持续提升工作效率。

在每一个疑难点中，本书先分析再给出建议，对疑难点涉及的技术点单独给出关键词，在避坑指南中对易出错的地方给出提示。同时，对于一些需要结合特定场景的疑难点，给出了完整的参考代码。为便于读者随手做笔记，在疑难点的最后预留了空白处。关于文中使用的全部参考代码，可以扫描本书序言中的二维码，或者关注微信公众号"路科验证"，在输入栏输入"芯片验证调试手册"即可获得参考代码的下载链接。

第 2 章
SystemVerilog 疑难点集合

2.1 数据使用

SystemVerilog 引入的组合型数组和非组合型数组在存储和赋值方面均有差异。对于各种数组类型的访问和操作，也有容易发生错误的地方。将变量的生命周期分为静态的和动态的，也会影响程序在仿真时的行为。在参数方面，应理解参数对设计结构和验证环境的影响，清楚了解在编译和仿真阶段它们各自的生效方式。接下来我们将提出若干数据使用相关的疑难点，并展开讨论。

2.1.1 组合型数组和非组合型数组怎么区分？

组合型数组即 packed array，非组合型数组即 unpacked array。数组维度声明在数组变量名左侧的为组合型数组，例如 *byte [3:0] [1:0] array1* 为组合型 4*2 二维数组。数组维度声明在数组变量名右侧的为非组合型数组，例如 *byte array2 [5:0][7:0]* 为非组合型 6*8 二维数组。混合型数组 *byte[3:0] [1:0] array3 [5:0][7:0]* 为 6*8*4*2 的四维混合数组。

关键词：
packed　组合型，unpacked　非组合型

避坑指南：
数组维度从高到低是先看数组名右侧（从左到右），再看数组名左侧（从左到右）。

阅读手记：

2.1.2 组合型数组和非组合型数组如何做赋值？

组合型数组之间赋值时可以将不同维度、不同元素数量的数组直接赋值（注意，位宽默认补全或者截取）；非组合型数组之间赋值有严格要求，须维度相同，且各维度元素数量须相

等，满足该条件的两个数组可直接进行赋值，但如果不满足该条件，则只能对数组中的元素逐一赋值。

关键词：

packed 组合型，unpacked 非组合型，赋值

避坑指南：

组合型数组可直接赋值；非组合型数组直接赋值要求繁多，逐一赋值不易出错。

参考代码： sv_array_packed_unpacked.sv

```
module tb;
  bit [3:0][7:0] arr_mixed [1:0];
  logic [15:0] vec16;
  logic [31:0] vec32;
  logic [63:0] vec64;
  bit arr_unpacked [1:0][3:0][7:0];
  bit arr_unpacked_d2 [4][8];
  bit [1:0][3:0][7:0] arr_packed;

  initial begin: arr_assign
    $display("packed vector array assignment");
    vec16 = 16'hFFFF;
    xdisplay("vec16", vec16);
    vec16 = '1;
    xdisplay("vec16", vec16);
    vec16 = {4{4'hF}};
    xdisplay("vec16", vec16);
    vec32 = vec16;
    xdisplay("vec32", vec32);
    vec32 = 'hFF;
    xdisplay("vec32", vec32);
    vec64 = {vec32, vec32};
    xdisplay("vec64", vec64);

    $display("packed array slice assignment");
    arr_mixed[0] = vec32;
    xdisplay("arr_mixed[0]", arr_mixed[0]);
    arr_mixed[1] = vec64;
    xdisplay("arr_mixed[1]", arr_mixed[1]);
    arr_packed[0] = arr_mixed[1];
    xdisplay("arr_packed[0]", arr_packed[0]);
    // Errors below
    // arr_packed = arr_mixed;
    // arr_unpacked = arr_mixed;

    $display("unpacked array slice assignment");
    //arr_unpacked_d2 =  '{0:'{0:0, 1:0, default:1}, 1:'{0:1, 1:1,
```

```
default:0}, default:'{default:1}};
    arr_unpacked_d2 = '{'{0:0, 1:0, default:1},
                       '{0:1, 1:1, default:0},
                       '{0:0, 1:0, default:1},
                       '{0:1, 1:1, default:0}
                      };
    arr_unpacked_d2[0] = '{default:1};
    arr_unpacked[0] = arr_unpacked_d2;
    $finish();
  end

  function void xdisplay(string s, logic[63:0] val);
    $display("%s = 'h%x", s, val);
  endfunction
endmodule
```

仿真结果:

```
packed vector array assignment
vec16 = 'h000000000000ffff
vec16 = 'h000000000000ffff
vec16 = 'h000000000000ffff
vec32 = 'h000000000000ffff
vec32 = 'h00000000000000ff
vec64 = 'h000000ff000000ff
packed array slice assignment
arr_mixed[0] = 'h00000000000000ff
arr_mixed[1] = 'h00000000000000ff
arr_packed[0] = 'h00000000000000ff
unpacked array slice assignment
```

阅读手记:

2.1.3 在使用 enum 或 struct 时添加 typedef 与否的差别是什么？

如果不添加 typedef，例如 *enum {NO, YES} bool*，那么 bool 为枚举类型"变量"，而"*enum {NO, YES}*"按照"匿名类型"(anonymous type) 来理解；如果添加 typedef，例如 *typedef enum {NO, YES} bool_t*，那么 bool_t 为枚举类型，即通过 typedef 将匿名类型"显式"(explicitly) 定义为 bool_t，并且接下来可重复利用 bool_t 来声明多个变量。

关键词:

enum，struct，typedef

避坑指南：

对于重复使用的 enum 或 struct 定义，默认添加 typedef，先定义类型再利用该类型去声明变量。

参考代码：sv_enum_assignment.sv

```
module tb;
  typedef enum {RED=1, GREEN, BLUE, YELLOW} color_t;
  int val;
  color_t clr;
  initial begin
    $display("clr = %s OR %0d", clr, clr);
    clr = GREEN;
    val = clr;
    $display("clr = %s, val = %0d", clr, val);
    // legal value and type conversion
    val = 4;
    clr = color_t'(val);
    $display("clr = %s, val = %0d", clr, val);
    // illegal value and type conversion
    val = 6;
    clr = color_t'(val);
    $display("clr = %s, val = %0d", clr, val);
    // $cast(TGT, SRC)
    val = 6;
    if(!$cast(clr, val))
      $error("casting failure while val = %0d", val);
    else
      $display("clr = %s, val = %0d", clr, val);
    $finish();
  end
endmodule
```

仿真结果：

```
clr =  OR 0
clr = GREEN, val = 2
clr = YELLOW, val = 4
clr = , val = 6
Error: "sv_enum_assignment.sv", 21: tb: at time 0 ps
casting failure while val = 6
```

阅读手记：

2.1.4 什么是静态变量和动态变量？

module、interface 和 package 中的变量和方法都是（默认为）静态的（static），不需要显性声明它们的静态属性，即它们的"生命"在仿真加载后即存在并且伴随至仿真结束；class 中的变量和方法则恰好相反，都是（默认为）动态的（automatic），只有在创建该对象的同时才会为其成员开辟空间，而一旦该对象被销毁，其成员的空间也将一同释放。

关键词：

static，automatic

避坑指南：

为测试平台准备的 module、interface 和 package 中的方法，请"习惯性"地为它们添加 automatic 描述符，这可以避免很多潜在的麻烦。

阅读手记：

2.1.5 struct 和 struct packed 区别在哪里？

非组合型结构体（默认类型）和组合型结构体在赋值方面的差异与"非组合型数组和组合数组"保持一致，且非组合型结构体占据空间要比组合型结构体更多。组合型结构体可以将所有数据直接赋值给一个向量，这方便了数据结构的存储，以及作为比特流（bit stream）传输时的打包和解包之用。

关键词：

struct，struct packed

避坑指南：

如果要做随机化，这两种类型的变量均应声明为 rand，且可对非组合型结构体成员单独指定 rand 属性（即可单独随机化成员），而不能对组合型结构体成员单独指定 rand 属性（只可以做整体结构随机化）。

参考代码： sv_struct_unpacked_diff_packed.sv

```
module tb;
  typedef struct {
    rand bit[7:0] m1[4];
    int m2;
    byte m3[4];
  } data1_t;

  typedef struct packed {
    bit [7:0][3:0] m1;
    int unsigned m2;
```

```
    bit[31:0] m3;
  } data2_t;

  class packet;
    rand data1_t d1;
    rand data2_t d2;
    rand bit[95:0] vec;
    constraint cstr {vec == d2;};
  endclass

  initial begin
    packet p = new();
    void'(p.randomize());
    $display("p = %p", p);
  end
endmodule
```

仿真结果：

```
p = '{d1:'{m1:'{'ha3, 'h8f, 'h11, 'h8a} , m2:0, m3:'{0, 0, 0, 0} },
      d2:'{m1:'h97ae37b4, m2:'h6ec8f6d, m3:'heda662ac},
      vec:'h97ae37b406ec8f6deda662ac}
```

阅读手记：

2.1.6 关联数组的散列存储表示什么？

关联数组的"散列存储"有三个意思。第一，关联数组在声明时，无须声明大小（与队列类似）；第二，关联数组的索引类型可以为任何类型，且对索引值大小、顺序不作要求；第三，关联数组的元素在存储空间中也是非连续存放的。

关键词：

associated array　关联数组，散列存储

避坑指南：

在使用 foreach() 对关联数组的索引进行逐一访问时，索引值的访问顺序按照默认的数字排列顺序，或者按照字母排列顺序有序排列。

阅读手记：

2.1.7 如何将队列插入到另外一个队列中？

在使用中容易出现这种不恰当的用法，*q1.insert(x, q2)*，即在 q1 的 index=x 处，插入队列 q2。在实际使用中，有的仿真器可以支持而有的仿真器并不能支持该种使用方法。IEEE-1800 SystemVerilog-2017 的语法要求 *queue::insert(index, element)* 的第二个参数应该为队列中的单个元素，而不是某个队列。

关键词：
queue 队列，insert 插入队列

避坑指南：
我们还可以通过'{}'拼接操作符实现队列的插入。

参考代码： sv_queue_insert.sv

```
module tb;
  int q1[$] = {0, 1, 2, 6};
  int q2[$] = {3, 4};
  initial begin
    q1.insert(3, 5); // '5' is an element type q1 = {0, 1, 2, 5, 6}
    //q1.insert(3, q2); // unavailble operation due to q2 is not an element
    q1 = {q1[0:2], q2, q1[3:$]}; // availble insertion
    $display("%p", q1);
  end
endmodule
```

仿真结果：

'{0, 1, 2, 3, 4, 5, 6}

阅读手记：

2.1.8 队列在赋值时使用操作符{}，那么它属于组合型吗？

这是一个好问题，它把赋值方式和数组的类型结合在一起考虑。组合型和非组合型的概念，是相对于定长数组而言的，但是不能将队列等同为"组合型数组"，或是连续存储的。采用{}而不是'{}赋值，只是表明队列的赋值方式与组合型数组的赋值方式相同。

如果将一个向量赋值给队列，那么编译器会提示编译错误，即赋值等号两端的数据类型不匹配。这其实也告诉我们，不能将队列简单视为组合型数组，而且队列的数据也不是连续存储的。

关键词：
queue，packed array

参考代码：sv_queue_assignment.sv

```
module tb;
  bit[7:0] q[$];
  initial begin
    q = {8'hF, 8'h7, 8'hF, 8'h7};
    // compilation error:
    // type of source is incompatible with the target
    // q = 32'h0F070F07;
  end
endmodule
```

阅读手记：

2.1.9 数组的选取可以用两个变量作为索引边界吗？

实际上，数组索引的问题，跟索引变量是否为动态变量没有关系。在对数组索引时，应该避免的问题是用两个变量进行索引，例如 *bit[31:0] v1; int idx1 = 3, idx2 = 0;* 那么 *v1[idx1: idx2]* 在编译时，编译器不允许通过。但是可以选择一个变量作为某一个索引，而采取固定位宽的方式去选取某一个数组的片段（slice）。

关键词：
数组选取，array slice，index

避坑指南：
遗憾的是，目前没有好的办法可以直接利用两个变量去选择某个数组的片段，但是可以通过 "-=" "+=" 的片段选择符号，或 "<<" ">>" 的位移操作符号，间接达到选取数组片段的目的。

参考代码：sv_array_variable_slice.sv

```
module tb;
  bit [31:0] v1 = 32'h11223344;
  int idx1 = 31, idx2 = 24;
  initial begin
    // Error slice
    //$display("v1 high 1st byte is 8'h%0x", v1[idx1 : idx2]);
    // Error slice
    //$display("v1 high 1st byte is 8'h%0x", v1[idx1 : idx1-7]);
    // Correct slice
    $display("v1 high 1st byte is 8'h%0x", v1[idx1 -: 8]);
    // Correct slice
```

```
        $display("v1 high 1st byte is 8'h%0x", v1 >> idx2);
    end
endmodule
```

仿真结果：

```
v1 high 1st byte is 8'h11
v1 high 1st byte is 8'h11
```

阅读手记：

2.1.10 parameter、localparam 和 const 有什么联系和差别？

相比于 localparam，parameter 可以在实例例化时从外部修改。parameter 值的修改，可以在实例例化时通过#以 order list 或 name list 的方式来完成，也可以通过 defparam 赋值语句来实现。

localparam 是局部参数，无法再次修改，自然也不能通过在实例例化时从外部修改。localparam 在赋值时可以引用 parameter。

parameter 与 localparam 都是在 elaboration 时确定的，而 const 常量则是在仿真时确定的。与 localparam 类似的地方在于，const 一旦定义和初始化也就无法再修改它的值。

关键词：

parameter，localparam，const

避坑指南：

parameter 的值除了在设计代码中给定，也可以在编译时通过添加与参数修改有关的选项来给定（也可以覆盖设计代码中的给定 parameter 值），但它们的特征都会在 elaboration 阶段确定下来，与 parameter 有关的设计结构、数据位宽、逻辑行为等也因此都会在编译阶段确定下来。

const 常量虽然在代码中不能再次对其进行赋值，但仍然可以通过在仿真时的 force 处理来修改它的值，继而影响与其相关的逻辑行为。需要注意的是，不能通过 force 来修改 parameter 和 localparam 的数值。

参考代码：sv_para_localpara_const_diff.sv

```
module mod_a;
    parameter int data_width_p = 32;
    parameter int addr_width_p = 16;
    localparam int data_bytes_lp = data_width_p >> 3;
    localparam int addr_bytes_lp = addr_width_p >> 3;
endmodule
```

```
module tb;
  localparam int data_width_lp = 64;
  localparam int addr_width_lp = 32;
  // declaration with initial value
  const int data_width = 64;
  // declaration with value assigned later (only once)
  const int addr_width = 32;

  mod_a #(data_width_lp, addr_width_lp) m1();
  mod_a m2();
  defparam m2.data_width_p = 128;
  defparam m2.addr_width_p = 64;

  task automatic init();
    localparam int data_val_lp = 10;
    const int data = data_val_lp;
  endtask

  initial begin
    init();
  end
endmodule
```

阅读手记：

2.1.11 多维数组的声明和使用，哪种方式更合适呢？

一般我们建议多维数组的直接声明不超过 3 个维度（例如，*int arr_fix_d1 [2][3][4]*）。在声明多维数组时，可以伴随有 rand 属性，或将数组类型声明为动态数组等常见操作。

如果是多维动态数组，一种简单的方法是直接声明多维动态数组。在要求随机属性时，可以在约束中对多维数组的每个维度进行约束。

对于复杂的结构，可以先使用结构体来封装，再进行逐级嵌套，而在上层约束中对整个结构体构成的多维数组进行约束。如果不仅数据成员复杂，而且每一级的数据成员之间也有较复杂的约束，那么可以先使用类来进行数据封装，并在该层次中声明约束，而后进行逐层封装。

关键词：

multi-dimension，array，class，struct

避坑指南：

对于通过类来逐层封装多维数组的情况，不但要注意在句柄数组前使用 rand 来描述，还要考虑在每一级通过函数 post_randomize()对动态数组的句柄元素进行对应的实例创建和句柄复制。

上面这一步很重要，因为上层类的随机化只会随机化数组的大小，而不会创建下层类的实例。要实现这一目的，可以通过 post_randomize()函数来完成（其伴随着对 randomize()函数的调用）。

参考代码：sv_multi_dimension_array.sv

```
module tb;
  typedef struct {
    rand int d0[];
  } ds0_t;
  typedef struct {
    rand ds0_t d1[];
  } ds1_t;
  class data_pool;
    int arr_dyn_d1 [][];
    rand int arr_dyn_d2 [][];
    rand int arr_dyn_d3 [][][];
    rand ds1_t arr_dyn_d4 [];
    int arr_fix_d1 [2][3][4];
    rand int arr_fix_d2 [2][3][4];
    constraint cstr {
      // constraint to dynnamic array with 2-dimesion
      arr_dyn_d2.size() == 2;
      foreach(arr_dyn_d2[i]) arr_dyn_d2[i].size() == 3;

      // constraint to dynnamic array with 3-dimesion
      arr_dyn_d3.size() inside {[1:2]};
      foreach(arr_dyn_d3[i]) {
        arr_dyn_d3[i].size() inside {[2:3]};
        foreach(arr_dyn_d3[i][j]) arr_dyn_d3[i][j].size() inside {[3:4]};
      }
      // constraint to dynnamic array with 3-dimesion
      arr_dyn_d4.size() inside {[1:2]};
      foreach(arr_dyn_d4[i]) {
        arr_dyn_d4[i].d1.size() inside {[2:3]};
        foreach(arr_dyn_d4[i].d1[j])
          arr_dyn_d4[i].d1[j].d0.size() inside {[3:4]};
      }
    }
  endclass

  class dim0;
    rand int d0[];
    constraint cstr {
      d0.size() inside {[3:4]};
```

```
    }
  endclass
  class dim1;
    rand dim0 d1[];
    constraint cstr {
      d1.size() inside {[2:3]};
    }
    function void post_randomize();
      foreach(d1[i]) begin
        d1[i] = new();
        void'(d1[i].randomize());
      end
    endfunction
  endclass
  class dim2;
    rand dim1 d2[];
    constraint cstr {
      d2.size() inside {[1:2]};
    }
    function void post_randomize();
      foreach(d2[i]) begin
        d2[i] = new();
        void'(d2[i].randomize());
      end
    endfunction
  endclass

  initial begin
    data_pool dp = new();
    dim2 dm = new();
    assert(dp.randomize())
    else $error("data pool randomization failed!");
    assert(dm.randomize())
    else $error("dim object randomization failed!");
    $finish();
  end
endmodule
```

阅读手记：

2.2 操作符使用

对于 SystemVerilog 中的操作符，在每个独立场景中理解它们并不困难，而可能让用户困惑的一点在于，相同的操作符在不同场景中起不同的作用。这就需要记忆、分辨同一个操作符的不同使用场景和作用。同时，验证环境中存在硬件范围和软件范围，在这些不同的范围中使用什么样的数据类型、适配什么样的操作符，也都需要掌握。由于 logic 相关数据类型可以表达 X、Z 值，那么围绕着它们的操作符也要注意在 X、Z 值出现时的执行结果。接下来，我们将围绕操作符使用的疑难点展开论述。

2.2.1 { }操作符的使用场景有哪些？

第一，可作为并置操作符"连接"一些向量或字符串，例如{a, b}；第二，可作为复制操作符，例如{4{w}}中内嵌的一对{ }；第三，对组合型数组赋值时，可采用{ }，对非组合型数组赋值时，可采用'{ }（注意单撇号）；第四，在描述覆盖点和仓时使用{ }；第五，在描述约束时使用{ }；第六，在定义 enum、struct 时，使用{ }。其他场景不在这里一一列举，请读者试着在阅读手记处列出。

关键词：
curly bracket　花括号，{ }

避坑指南：
SV 中应该使用 begin...end 的字段切勿用{ }替代。

参考代码： sv_curly_brace_usage.sv

```
module tb;
  typedef enum {WRITE, READ, IDLE} op_t;
  class rkv_packet;
    rand int unsigned data [];
    rand byte unsigned addr;
    rand op_t op;
    constraint cstr {
      data.size() inside {[3:5]};
    }
    covergroup cg;
      CP_ADDR: coverpoint addr[1:0] {
        bins addr_00 = {2'b00};
        bins addr_01 = {2'b01};
        bins addr_10 = {2'b10};
        bins addr_11 = {2'b11};
      }
    endgroup
    function new();
```

```
      cg = new();
    endfunction
  endclass

  int unsigned darr[];
  rkv_packet pkt;
  initial begin
    darr = new[2];
    darr = '{2{32'h11}};
    darr = {darr, darr};
    pkt = new();
    if(!pkt.randomize() with {
      data.size == darr.size;
      foreach(data[i]) data[i] == darr[i];
      addr == {6'h30, 2'h0};
      })
      $error("rkv_packet randomization failure!");
    $finish();
  end
endmodule
```

阅读手记:

2.2.2　条件赋值符 ?: 和条件语句 if-else 的执行结果一致吗？

多数情况下二者的执行结果是一致的。但要注意，避免让条件判断结果出现 X 值，否则条件赋值符的结果会让你吃惊的（也会让你为此而调试查找问题）。原因在于，如果 ?: 赋值中的条件结果为 X 值，那么变量被赋值时很可能有 X 值出现，具体原因可参考 IEEE-1800 SystemVerilog-2017 标准 11.4.11 节的内容。

关键词：

?:, if-else

避坑指南：

无论上述哪种条件执行方式，尽量使用 case equality operators '==='、'!==', 而不使用 logical equality operators '=='、'!=', 避免条件本身得出 X 结果。

参考代码： sv_condition_diff_if.sv

```
module tb;
  logic cond0 = 1'bx;
  logic [1:0] val0 = 2'b01, val1 = 2'b10;
```

```
    logic [1:0] res;

    initial begin
      if(cond0 != 0)
        res = val0;
      else
        res = val1;
      $display("if(cond0 != 0) result in res = 'b%0b", res);
    end

    initial begin
      if(cond0 !== 0)
        res = val0;
      else
        res = val1;
      $display("if(cond0 !== 0) result in res = 'b%0b", res);
    end

    initial begin
      res = (cond0 != 0) ? val0 : val1;
      $display("(cond0 != 0) ?: result in res = 'b%0b", res);
    end

    initial begin
      res = (cond0 !== 0) ? val0 : val1;
      $display("(cond0 !== 0) ?: result in res = 'b%0b", res);
    end
endmodule
```

仿真结果：

```
if(cond0 != 0) result in res = 'b10
if(cond0 !== 0) result in res = 'b1
(cond0 != 0) ?: result in res = 'bxx
(cond0 !== 0) ?: result in res = 'b1
```

阅读手记：

2.2.3 if 和 iff 的应用场景分别有哪些？

这两个关键词的应用场景没有重叠。if 应用于条件判断，可以在 if-else 条件语句中，也可以在约束或断言属性中设置条件。iff 往往伴随着事件（event），其作为事件等待的"附加条件"，无论是在@(EVENT iff COND)时还是在covergroup、coverpoint 定义时，都可以添加 iff，例如@(posedge clk iff rstn === 1)。

关键词：
if，iff，条件

避坑指南：
凡是与事件等待有关（@EVENT）的附加条件，都应该使用 iff。

阅读手记：

2.2.4 使用 foreach 在轮循数组时按照什么顺序呢？

对于队列或者动态数组，foreach 循环时的 index 索引从 0 开始；对于定长数组，这取决于数组声明的索引顺序，例如，*int arr1[3:1]*、*int arr2[1:3]*，arr1 的索引值从 3 到 1，arr2 的索引值从 1 到 3（不一定从 0 开始或结束）；对于关联数组，最常见的 int 类型索引值会按照默认的排列顺序从小到大开始，string 类型的索引值也按照默认的顺序（a~z）排列。

关键词：
foreach，index，索引顺序

避坑指南：
应该注意定长数组和关联数组的索引方式，尤其是关联数组的不同索引数据类型，其排列方式有所不同。

参考代码： sv_foreach_loop_order.sv

```
module tb;
  int arr1 [2:0];
  int arr2 [1:3];
  int arr3 [int] = '{'h0:10, 'h4:20, 'h8:30};
  int arr4 [string] = '{"a":10, "b":20, "aa":30};
  initial foreach(arr1[i]) $display("arr1 index order %0d", i);
  initial foreach(arr2[i]) $display("arr2 index order %0d", i);
  initial foreach(arr3[i]) $display("arr3 index order %0d", i);
  initial foreach(arr4[i]) $display("arr4 index order %s", i);
endmodule
```

仿真结果：

```
arr1 index order 2
arr1 index order 1
```

```
arr1 index order 0
arr2 index order 1
arr2 index order 2
arr2 index order 3
arr3 index order 0
arr3 index order 4
arr3 index order 8
arr4 index order a
arr4 index order aa
arr4 index order b
```

阅读手记：

2.2.5 运算符的优先级是否有必要记忆呢？

即便想记住，恐怕也不容易吧。除了常见的加减乘除，在硬件运算逻辑里面容易忽略的一些运算符（操作符）的优先级，包括等值操作符==、!=、===、!==，移位操作符 <<、>>，按位操作符|、^以及逻辑操作符&&、||等。如果在运算时，以上这些运算符发生了混用，那么"强烈"建议大家使用小括号来表明运算意图！这可以避免很多因疏忽而带来的运算结果错误。甚至更糟糕的是，一时的疏忽大意还会让你在调试过程中困扰不已，直到某个时刻大脑迎来了光，才会茅塞顿开呢！

关键词：
operator 运算符（操作符），precedence 优先级

避坑指南：
遇到移位操作符、等值操作符、按位操作符等，建议使用小括号表明运算意图。例如，$A + B << 4$，并不表示 A+B 以后再左移 4 位，而实际上是 B 先左移 4 位再与 A 相加；又比如 $A \wedge B !== C$，并不表示 A^B 运算后再与 C 判断等值，而是 B 先与 C 判断等值，待得出其逻辑结果 1 或 0 后，再与 A 进行异或运算。

这里附上 IEEE-1800 SystemVerilog-2017 标准手册关于运算符的优先级表格。

运算符	可结合性	优先级
() [] :: .	左	最高
+ - ! ~ & ~& | ~| ^ ~^ ^~ ++ -- (unary)		
**	左	
* / %	左	
+ - (binary)	左	
<< >> <<< >>>	左	
< <= > >= inside dist	左	
== != === !== ==? !=?	左	
& (binary)	左	
^ ~^ ^~ (binary)	左	
| (binary)	左	
&&	左	
||	左	
?: (conditional operator)	右	
–> <–>	右	
= += -= *= /= %= &= ^= |= <<= >>= <<<= >>>= := :/ <=	无	
{} {{}}	关联性	最低

阅读手记：

2.2.6 assign 连续赋值可以赋值给 logic（var）类型吗？

首先要明白，在 SV 中，类型可以分为线网（net）和变量（variable）。线网的赋值设定与 Verilog 的要求相同，即线网赋值需要使用连续赋值语句（assign），而不应出现在过程块（initial、always）中；相比于线网驱动的限制，变量（var）类型的驱动要求就没那么多了，如 logic [3:0] a，该变量默认类型是 var（变量），对它可以使用连续赋值或过程赋值。

关键词：

assign，logic，var，net，wire

避坑指南：

简而言之，可以在 testbench（module）中的数值存储和线网连接，多数情况下使用 logic 类型变量，而很少有只能使用 wire 的情况。那么什么时候需要使用 wire 呢？例如，当出现多于一个驱动源的时候，或设计模块端口是双向（inout）的时候，都需要使用 wire 来完成线网连接。

参考代码：sv_assign_wire_var.sv

```
module tb;
  wire w1;
  logic r1; // logic (var)
  assign w1 = 1'b1;
  assign r1 = 1'b0;

  wire logic w2;
  var logic r2;
  assign w2 = 1'b1;
  assign r2 = 1'b0;
  initial begin
    #0;
    $display("w1 = %b, r1 = %b", w1, r1);
    $display("w2 = %b, r2 = %b", w2, r2);
  end
endmodule
```

仿真结果：

```
w1 = 1, r1 = 0
w2 = 1, r2 = 0
```

阅读手记：

2.2.7　:: 和 . 这两个符号使用起来有哪些区别？

:: 为域的索引符（scope resolution operator），即从一个域中索引某些类型、变量或方法。比如，从某个 package 或某个类中索引某个类型、变量或方法，又比如我们在调用 *std::randomize()* 函数时也是从 std 库（预定义）来引用 randomize() 函数。

. 为层次化索引方式（hierarchy reference），在 Verilog 设计层次中，可以通过.来索引绝对路径或相对路径下的变量，在 SV 的验证环境中可以通过.来索引验证结构中的某个子一级实例或它的成员，也可以通过.来索引结构体变量的成员。此外，. 还可以用来调用目标中的方法。

关键词：
::，域索引，.，层次索引

避坑指南：
在使用:: 时，所调用方法或变量应该为静态类型（不支持使用:: 索引动态类型）；在使用. 时，所调用的路径层次（无论是硬件层次还是软件层次）都应该存在。否则，不存在的硬件层次在链接时（elaboration）会发生错误，而不存在的软件层次在链接时或在运行时会发生错误。

阅读手记：

2.3 模块、接口与方法

模块与接口是与验证环境通信的三大类型。模块与接口中的变量、方法由于默认是静态属性，所以需要同类中的成员做区别。模块与接口的过程块执行顺序、彼此的例化关系也需要注意。接口是硬件范围与软件范围之间的数据交换中介，要掌握它在硬件范围和软件范围中的使用方式。时钟块对于接口是一个重要的部分。理解时钟块的驱动、采样行为表现，对于正确使用时钟块有重要意义。接下来我们将围绕模块、接口与方法，展开论述一些疑难点。

2.3.1 module 中的方法在声明时是否要添加 automatic？

首先需要清楚，在 module、interface 中声明的无论是方法还是变量，默认都是静态属性（static），这意味着 module 中的 task、function 默认是静态方法。同时，这些静态方法中的变量默认也是静态的，因此在调用它们时会遇到静态变量的存储共享，可能会出现不同线程在同一时间调用同一方法，继而引起结果干扰的问题。

关键词：

module，interface，task，function，dynamic，static

避坑指南：

module、interface 中的方法在声明时建议添加 automatic，避免可能会出现的存储共享的问题。

参考代码： sv_static_and_automatic.sv

```
package rkv_pkg;
  class rkv_packet;
    int data[];
    static local int _id;
    function new();
      _id++;
    endfunction
    static function int get_id();
      return _id;
    endfunction
  endclass
endpackage

module tb;
  import rkv_pkg::*;
  rkv_packet pkts[];
  function int static_increment(int n = 1);
    static int value = 0;
    value += n;
    $display("static_increment result is %0d with n = %0d", value, n);
```

```
    return value;
  endfunction

  function automatic int auto_increment(int n = 1);
    int value = 0;
    value += n;
    $display("auto_increment result is %0d with n = %0d", value, n);
    return value;
  endfunction

  initial begin
    pkts = new[3];
    foreach(pkts[i]) begin
      if(i>0)
        $display("pkts[%0d] id is %0d", i-1, rkv_packet::get_id());
      pkts[i] = new();
      $display("pkts[%0d] id is %0d", i, pkts[i].get_id());
    end
    repeat(3) begin
      static_increment($urandom_range(1, 4));
      auto_increment($urandom_range(1, 4));
    end
  end
endmodule
```

仿真结果:

```
pkts[0] id is 1
pkts[0] id is 1
pkts[1] id is 2
pkts[1] id is 2
pkts[2] id is 3
static_increment result is 3 with n = 3
auto_increment result is 1 with n = 1
static_increment result is 5 with n = 2
auto_increment result is 3 with n = 3
static_increment result is 9 with n = 4
auto_increment result is 4 with n = 4
```

阅读手记:

2.3.2 interface 在何处需要使用 virtual 来声明呢？

首先记住 interface 在验证环境中的作用，那就是充当软件环境（验证环境）和硬件环境（RTL 设计）交互的媒介。interface 在 TB（module）中例化时，需要参考 module 实例化的方式，因此在 module 中它是"硬件属性"。验证环境中的各个验证组件均需要通过 interface 去访问接口信号，因此在 class 中它是"软件属性"，并且需要添加 virtual 表示接口句柄。

关键词：

interface 接口，interface handle 接口句柄

避坑指南：

module 中可以声明 virtual interface（但用处不多，更多的是通过层次化引用接口实例去访问接口成员），在 class 中只能使用 virtual interface（而无法例化接口实例）。

参考代码： sv_interface_virtual_reference.sv

```
interface intf;
  logic a;
  logic b;
endinterface

class driver;
  virtual intf vif;
endclass

module tb;
  intf ifi0();
  intf ifi1();
  intf ifis[2] ();
  virtual intf vifs[4];
  driver drv;
  initial begin
    vifs = '{ifi0, ifi1, ifis[0], ifis[1]};
    vifs[0].a = 1'b1;
    drv = new();
    drv.vif = ifi1;
    drv.vif.b = 1'b0;
    $finish();
  end
endmodule
```

阅读手记：

2.3.3 initial 和 always 的执行顺序是否与代码位置有关？

always 过程语句块一般在综合逻辑中，initial 过程语句块一般在初始化激励序列中，它们彼此之间都是"并行"的，这是从模拟硬件执行角度来理解"并行"。与此同时，仿真器即便要处理这些过程块，也需要从软件语句执行着手。这意味着在同一个仿真时刻去执行的（并行）语句块，需要由仿真器安排它们的执行顺序，这是从仿真软件执行角度来理解"顺序"。

关键词：
执行顺序，initial，always，parallel，order

避坑指南：
不对多个 initial 过程块的执行顺序做假设。如果要按照顺序执行，可将逻辑放置在同一个 initial 过程块，或使用 event 实现线程间的同步。

参考代码： sv_parallel_initial_process.sv

```
module tb;
  event sync_e1, sync_e2;
  bit flag1, flag2;
  initial begin : ini_pro1
    $display("ini_proc1 triggered sync event sync_e1");
    -> sync_e1;
    wait(sync_e2.triggered);
    $display("ini_proc1 got sync event sync_e2");
    flag1 = 1;
  end
  initial begin : ini_pro2
    wait(sync_e1.triggered);
    $display("ini_proc2 got sync event sync_e1");
    -> sync_e2;
    $display("ini_proc2 triggered sync event sync_e2");
    flag2 = 1;
  end
  initial begin
    wait({flag1, flag2} == 2'b11);
    $display("ini_proc1 and ini_proc2 finished!");
    $finish();
  end
endmodule
```

仿真结果：

```
ini_proc1 triggered sync event sync_e1
ini_proc2 got sync event sync_e1
ini_proc2 triggered sync event sync_e2
ini_proc1 got sync event sync_e2
ini_proc1 and ini_proc2 finished!
```

阅读手记：

2.3.4 interface 的 modport 和 clocking block 如何使用？

这两个概念需要独立看待。modport 是将 interface 中的信号列表按类分簇（grouping），便于模块之间、模块与 TB 之间的连线管理；clocking block 并非 interface 的专属产物，不过多见于 interface 使用它进行数据信号采样或驱动，继而有效避免 delta cycle 问题，并且可以通过波形上的可见延迟帮助理解仿真时序。

关键词：

interface，modport，clocking block

避坑指南：

modport 和 clocking block 在连接时方向不要混淆，在它们中只需要对若干已声明的信号再次声明方向即可。

参考代码： sv_interface_clocking_modport.sv

```
interface rkv_intf(input logic clk);
  logic req, grt;
  logic [7:0] addr, data;
  clocking cb @(posedge clk);
    input grt;
    output req, addr;
    inout data;
  endclocking
  modport mod_dut (
    input clk, req, addr, // async dut modport
    output grt,
    inout data
  );
  modport mod_ck_tb (clocking cb); // synchronous testbench modport
  modport mod_tb (
    input grt, // async testbench modport
    output req, addr,
    inout data
  );
endinterface
```

阅读手记：

2.3.5 module 和 interface 之间可以相互例化吗？

module 可以例化 module，也可以例化 interface；interface 可以例化 interface，但是不可以例化 module。就硬件实现逻辑而言，module 之间的嵌套、interface 之间的嵌套以及 module 嵌套 interface 都符合设计理念，而 interface 不需要（也不应该）例化 module。

关键词：

module，interface，instance

避坑指南：

interface 可以例化 interface，但无法例化 module。

参考代码： sv_module_interface_instance.sv

```
interface rkv_master_intf;
  int master_id;
endinterface

interface rkv_slave_intf;
  int slave_id;
endinterface

interface rkv_system_intf #(int master_num = 1, int slave_num = 1);
  rkv_master_intf master_ifs[master_num] ();
  rkv_slave_intf  slave_ifs[slave_num] ();
endinterface

module master #(int master_id = 0) (rkv_master_intf intf);
  initial intf.master_id = master_id;
endmodule

module slave #(int slave_id = 0) (rkv_slave_intf intf);
  initial intf.slave_id = slave_id;
endmodule

module top;
  master #(0) mst0 (sys_intf.master_ifs[0]);
  master #(1) mst1 (sys_intf.master_ifs[1]);
  slave  #(0) slv0 (sys_intf.slave_ifs[0]);
  slave  #(1) slv1 (sys_intf.slave_ifs[1]);
```

```
  rkv_system_intf #(.master_num(2), .slave_num(2)) sys_intf();
endmodule
```

阅读手记：

2.3.6 方法的参数默认方向该如何辨别？

对于 function、task，如果其参数的方向未声明，那么它的方向默认为 input 方向。如果声明了方向，那么该参数以及其后续的参数方向均相同。例如 *function void foo(A, B, output C, D)* 中，A 和 B 由于未声明方向，所以是 input，C 和 D 为声明后的方向，即 output。

关键词：

argument direction　参数方向，default direction　默认方向

避坑指南：

对每一个参数都应该声明方向（不易出错），且声明方向的同时记得添加数据类型，否则默认数据类型为 logic（1 位宽）。

参考代码： sv_method_args_direction.sv

```
module tb;
  function automatic void foo_a(input a, input b, output c);
    c = a + b;
  endfunction

  function automatic void foo_b(int a, int b, output int c, int d);
    c = a + b;
    d = c + 1;
  endfunction

  initial begin
    int val1, val2;
    foo_a(3, 2, val1);
    $display("foo_a(3, 2, val1) -> val1 = %0d", val1);
    foo_b(3, 2, val1, val2);
    $display("foo_b(3, 2, val1, val2) -> val1 = %0d, val2 = %0d",
             val1, val2);
  end
endmodule
```

仿真结果：

```
foo_a(3, 2, val1) -> val1 = 1
```

```
foo_b(3, 2, val1, val2) -> val1 = 5, val2 = 6
```

阅读手记：

2.3.7 return 的使用场景有哪些？

return 可在 function、task 中使用。在返回值为 void 的 function 或 task 中调用 return，会立即退出该方法；如果 function 返回值非 void，那么在退出 function 的同时还会返回数值。

关键词：

return，function，task

避坑指南：

在 task 中也可以使用 return 实现立即退出。

参考代码： sv_return_usage.sv

```
module tb;
  int val;
  function automatic int fun_foo(int a);
    return a;
  endfunction

  task automatic tsk_foo(ref int a);
    forever begin
      @a;
      if(a < 5) $display("a = %0d", a);
      else return;
    end
  endtask

  initial begin
    fun_foo(val);
    tsk_foo(val);
    $finish();
  end

  initial begin
    for(int i = 0; i< 10; i++)
      #1ns val = i;
  end
endmodule
```

仿真结果：
```
a = 1
a = 2
a = 3
a = 4
```

阅读手记：

2.3.8 task 与 function 的联系和差别在哪里？

task 和 function 均可在 module、interface、package 和 class 中定义，并实现一些功能。function 执行后须即刻返回，无法内置阻塞等待语句，在声明时需指定返回值（包括 void）。task 执行后无须即刻返回，可以内置阻塞等待语句（wait、#、@），返回数值只能依靠参数列表中的参数。

关键词：
task，function

避坑指南：
function 可以调用 function，但不建议调用 task（因为 task 可能内置阻塞等待语句）；task 既可以调用 task，也可以调用 function。

参考代码： sv_function_task.sv

```
module tb;
  function void func1();
  endfunction

  task t1();
    func1();
  endtask

  task t2();
    #1ns;
  endtask

  function void func2();
    // VCS compilation warning :
    // Warning-[TEIF] Task enabled inside a function
    t1();
    // VCS compilation error:
```

```
    // Error-[SV-DOSIF] Delay or synchronization in function
    t2();
  endfunction

  initial begin
    t1();
    func2();
  end
endmodule
```

阅读手记：

2.3.9 方法的参数默认值该如何使用？

方法的参数默认值有实用价值（不建议使用参数默认方向和默认数值类型）。在调用方法并传递参数过程中可按位置或按名称传递参数。在按位置传递参数的过程中，如果某个参数具备默认值且使用该值，那么需要为它留好"空的位置"，除非该默认参数的位置在参数列表的最后。

关键词：
argument 参数，default value 默认值，参数传递

避坑指南：
将所有带有默认值的参数声明均放在参数列表的最后，便于外部在调用方法时可以缺省参数。此外，如果参数带有默认值，也便于后期对这些方法进行参数扩充，而不至于对使用过这些方法的地方进行相应的方法调用形式（参数传递）的更新。

参考代码： sv_argument_default_val.sv

```
module tb;
  class rkv_packet;
    local int _data;
    function void set_data(input int val, input bit reverse = 0);
      _data = reverse ? val : (0 - val);
      $display("rkv_packet::set_data(.val(%0d), .reverse(%0d)) is called", val, reverse);
    endfunction
    function int get_data();
      return _data;
    endfunction
  endclass
  class user_packet extends rkv_packet;
    function void set_data(input int val,
```

```
                                        input bit reverse = 0,
                                        input bit double = 0);
      if(double) val = 2 * val;
      super.set_data(val, reverse);
      $display("user_packet::set_data(.val(%0d), .reverse(%0d), .double(%0d)) is called", val, reverse, double);
    endfunction
  endclass

  rkv_packet rp;
  user_packet up;
  initial begin
    rp = new();
    rp.set_data(100, -1);
    $display("rp.data is %0d", rp.get_data());
    up = new();
    up.set_data(200, -1);
    $display("up.data is %0d", up.get_data());
    $finish();
  end
endmodule
```

仿真结果：

```
rkv_packet::set_data(.val(100), .reverse(1)) is called
rp.data is 100
rkv_packet::set_data(.val(200), .reverse(1)) is called
user_packet::set_data(.val(200), .reverse(1), .double(0)) is called
up.data is 200
```

阅读手记：

2.3.10 方法中参数方向 inout 和 ref 有什么差别？

inout 方向会在方法调用中完成入口处由外部变量到形式参数的复制，以及在方法退出时由形式参数到外部变量的复制，即一共 2 次值的复制；ref 则是将外部变量本身传递进入，即不再发生形式参数的值的复制过程，ref 也可以理解为指针（reference）。

关键词：

argument direction　　形式参数方向，inout，ref

避坑指南：

如果要对某个外部变量进行持续"跟踪"，那么应该使用 ref 方向描述符，并且在 task 中对其进行跟踪。

参考代码：sv_args_inout_diff_ref.sv

```
module tb;
  function automatic void extend_array(inout int arr[]);
    arr = new[arr.size()+1] (arr);
  endfunction

  task automatic wait_array_size(ref int arr[], input int size);
    wait(arr.size() == size);
  endtask

  int arr[];

  initial begin : extend_proc
    for(int i=0; i<10; i++) begin
      extend_array(arr);
      arr[arr.size()-1] = i;
      #1ns;
    end
  end

  initial begin : wait_proc
    wait_array_size(arr, 5);
    $display("arr size reached 5 with members %p", arr);
    $finish();
  end
endmodule
```

仿真结果：

```
arr size reached 5 with members '{0, 1, 2, 3, 4}
```

阅读手记：

2.3.11 module 和 interface 中的变量声明必须放置在头部位置吗？

理论上，变量声明（reg、wire、var）可以放置在 module、interface 中的任意位置，但我们仍然建议将所有变量的声明放置于 module、interface 的头部，这是由于隐式变量声明（implicit variable declaration）的存在，即在端口声明、模块例化端口信号连接、连续赋值（assign）发生时，如果该变量未声明过，那么在编译时会隐式地为其指定一个变量（1 位 logic 类型），并且覆盖其后用户可能声明过的同名变量。

关键词：

variable declaration　变量声明，implicit variable declaration　隐式变量声明

避坑指南：

请将变量声明尽可能放置于 module、interface 的头部位置（或至少记得声明这些变量），同时可使用 linting 工具辅助检查，或仿真器编译时添加 linting 选项（例如，使用 VCS 时可添加编译选项+lint=all）。

参考代码： sv_module_implicit_variable.sv

```
module mod(
  input logic [3:0] in0
  ,input logic [3:0] in1
  ,output logic [3:0] out
  );
  assign out = in0 + in1;
endmodule

module tb;
  logic [3:0] val1;
  logic [3:0] val2;
  // imlicit 1-bit variable declared
  mod m0(val1, val2, out);
  // 4-bit variable explicitly declared but too late (no effect)
  logic [3:0] out;
  initial begin
    #1ns val1 = 'h1; val2 = 'h1;
    #1ns val1 = 'h2; val2 = 'h1;
    #1ns val1 = 'h3; val2 = 'h2;
    #1ns $finish();
  end
  always_comb begin
    $display("val1 = 'h%0x, val2 = 'h%0x, out = 'h%0x", val1, val2, out);
  end
endmodule
```

编译结果：

```
Warning-[PCWM-W] Port connection width mismatch
The following 1-bit expression is connected to 4-bit port "out" of module "mod", instance "m0".
```

仿真结果：

```
val1 = 'hx, val2 = 'hx, out = 'hx
val1 = 'h1, val2 = 'h1, out = 'h0
val1 = 'h2, val2 = 'h1, out = 'h1
val1 = 'h3, val2 = 'h2, out = 'h1
```

阅读手记：

2.3.12 如何例化和传递多个相同类型的接口？

除了做代码"复制粘贴"，还可以使用接口数组声明例化和 generate-endgenerate 语句块来完成接口的连接和传递。这种方式可以用于解决通过参数来控制例化接口数量的情况。

关键词：
interface instantiation　接口例化，interface assignment　接口传递

避坑指南：
仿真器不允许对接口数组使用变量进行索引访问，因此需要 genvar 变量以"文本平铺"的形式结合 for-loop 进行接口的连接和传递。

参考代码： sv_interface_instance_generate.sv

```
interface genif;
endinterface

module tb;
  import uvm_pkg::*;
  `include "uvm_macros.svh"
  localparam IFS_NUM = 4;

  // suggested way with generate-loop
  genvar i;
  genif ifs[IFS_NUM]();
  generate
    for(i = 0; i<IFS_NUM; i++) begin
      initial begin
        uvm_config_db#(virtual genif)::set(uvm_root::get(),
          $sformatf("uvm_test_top.*agents[%0d]*", i), "vif", ifs[i]);
      end
    end
  endgenerate

  // available solution but not flexible due to constant index
  initial begin
    uvm_config_db#(virtual genif)::set(uvm_root::get(),
      "uvm_test_top.*agents[0]*", "vif", ifs[0]);
    uvm_config_db#(virtual genif)::set(uvm_root::get(),
      "uvm_test_top.*agents[1]*", "vif", ifs[1]);
```

```
      uvm_config_db#(virtual genif)::set(uvm_root::get(),
        "uvm_test_top.*agents[2]*", "vif", ifs[2]);
      uvm_config_db#(virtual genif)::set(uvm_root::get(),
        "uvm_test_top.*agents[3]*", "vif", ifs[3]);
    end
  endmodule
```

阅读手记：

2.3.13 使用 while 和 forever 语句时需要注意什么？

新手在使用 while 和 forever 语句时，容易在其循环结构中检查硬件信号或等待变量值。但应格外小心，如果循环结构中没有延时处理语句（#，wait，@），那么仿真极易在该循环结构中"卡死"（hang up）。这是因为从仿真的角度而言，如果你等待的信号或变量会在后续时刻发生，那么缺少阻塞等待处理的循环将会在当前仿真时刻占用仿真资源保持 active 状态。在这时，仿真无法继续前进到下一个仿真时刻，从而出现"奇怪的仿真时刻静止"现象。

关键词：
while，forever，blocking wait，hang up

避坑指南：
在使用 while 和 forever 循环语句检查硬件信号或等待变量时，一定要使用阻塞等待处理。如果对变量变化时刻捕捉不严格，那么可以加大循环延时间隔，节省仿真资源。

参考代码： sv_while_forever_deadloop.sv

```
module tb;
  logic v1, v2;
  bit clk;
  initial forever #5ns clk <= !clk;
  initial #21ns v1 <= 1'b1;
  initial #28ns v2 <= 1'b1;
  initial begin
    fork
      while(1) begin
        if(v1 === 1'b1) begin
          $display("@%0t got v1 === 1'b1", $time);
          break;
        end
        // important to add delay
```

```
        #10ns;
      end
      forever begin
        if(v2 === 1'b1) begin
          $display("@%0t got v2 === 1'b1", $time);
          break;
        end
        // blocking statement added
        @(posedge clk);
      end
    join
    $display("@%0t blocking process exited", $time);
    $finish();
  end
endmodule
```

仿真结果：

```
@30000 got v1 === 1'b1
@35000 got v2 === 1'b1
@35000 blocking process exited
```

阅读手记：

2.3.14 系统函数和内建方法有什么区别？

系统函数（system function）指的是 SystemVerilog 预定义的方法（function、task），均以 $符号开头，例如 $display()、$finish()；内建方法（built-in method）指的是 SystemVerilog 针对某些类型，为其内建的方法，例如，数组所具备的方法 size()、delete()，或类所具备的方法 randomize() 等。

关键词：

system function   系统函数，built-in method   内建方法

避坑指南：

数组大小的内建方法是 size()，而信箱大小的内建方法是 num()。类的内建方法 randomize()、系统函数 $random() 和 std::randomize() 是不同的随机化方法。

阅读手记：

2.3.15 接口和模块的联系和差别是什么？

模块构成整个硬件的层次；模块和模块之间的通信除了在 Verilog 中通过端口进行，也可以通过接口（interface）进行，还可以利用同一个接口完成多个模块之间的数据交互。接口同时也是模块（DUT）和外部验证环境的交互媒介，可将激励信号作为 DUT 的输入端，也可以将 DUT 的端口、内部信号交由验证组件进行监测和检查。

关键词：

interface 接口，module 模块

避坑指南：

模块可以例化模块或接口；接口可以例化接口，但不能例化模块（从设计意义上，这样的规范是合理的）。

阅读手记：

2.3.16 program 和 module 的联系和差别是什么？

program 是 SystemVerilog（非 UVM 环境）为测试而准备的测试环境的"外壳"。因此，program 中不可以出现硬件行为语句，例如，always、module、interface 和其他 program 例化。program 对数据的采样发生在 reactive 阶段，即采样的是硬件变量在变化后的稳定数据，从而避免采样时可能发生的竞争问题（数据不稳定）。此外，program 也具备自动结束仿真的隐性方法，这也是 module 所不具备的。

关键词：

program，module，reactive，sampling

避坑指南：

有了 UVM 环境之后，program 的上述特点或优势均被 UVM 相对应的特性所取代，例如，数据驱动采样被时钟块取代、自动结束仿真被 objection raise/drop 取代。无论读者是否熟悉 program 特性，都不再需要在 UVM 环境中使用它。

阅读手记：

2.3.17　多个线程在仿真调度中是如何在不同的域之间执行和切换的？

SystemVerilog 的仿真在多数情况下是以时钟事件为驱动点的，也就是在时钟沿对时序逻辑进行赋值，时序逻辑继而会影响组合逻辑，并且接下来进一步扩散影响其他时序逻辑或组合逻辑。这种"病毒"式的扩散，需要在每一个事件点（时钟沿为多数情况、时间延迟为少数情况）的不同调度区域（Preponed、Active、Inactive、NBA、Observed、Reactive、Postponed）依次进出，多个线程之间会发生相互触发、反复进入某个调度区域的情况，直到"病毒"扩散稳定后，才会进入 postponed 区域，继而准备进入下一个时间片（time-slot）。

关键词：
time-slot　时间片，schedule　调度，region　区域

避坑指南：
对数据的采样都应该在 Reactive、Postponed、Preponed 区域，program 和时钟块都可以做到这一点，从而保证数据采样的稳定。

阅读手记：

2.3.18　时钟块在使用时需要注意哪些地方？

时钟块（clocking block）在验证环境的一侧时，可以在 driver、monitor、coverage model 等多处使用。在 driver 中使用时，既利用了时钟块的驱动延迟特性，也利用了时钟块的稳定采样特性。要注意的是，时钟块中的采样信号较接口中的原始信号晚 1 拍，这是由于时钟块本身就是基于时钟沿做了稳定采样后将信号的值保留下来的。因此在 driver 中，采用的接口中的信号和时钟块的信号之间有 1 拍的延迟。如果没有注意到这一点，那么在 driver 的时序实现过程中将会遇到麻烦，使用者会不清楚什么时候该使用时钟块的采样信号，什么时候又该直接使用接口中的信号。

采用时钟块在 driver 中做延迟驱动的目标是可以理解的，但有时为了能够及时捕捉到信号变化，在 driver 中会直接使用接口中的信号而不是时钟块的信号（因为时钟块的信号比接口中的相同信号要晚 1 拍）。这一顾虑，在 monitor、covergrage model 等其他组件中则没有那么明显，这是因为这些组件都属于 passive 角色，即本身不需要与接口信号互动（即从接收信号到反馈驱动），所以在这些组件中可以统一采用时钟块中的信号。由于这些信号是稳定采样

的，如果全部使用它们来解析时序，那么相当于整体时序的分析比实际信号时序延迟了 1 拍，而这 1 拍的整体延迟对于绝大多数协议时序而言是可以接受的（当然，也可以推断出来实际的时序事件发生的时间点）。

关键词：
clocking block，driver，monitor，coverage model

避坑指南：
在 driver 或 monitor 中实现时序驱动或采样时，不一定要求每个时序事件都基于时钟沿。为了满足时序，有时也需要模拟出组合时序行为，即在某个时钟沿内需要利用 wait、#、@等延时阻塞等待操作，去采样某些信号。这种实现方式当然也是可以的。验证环境中的组件在实现时序时，并不需要像设计代码那样去严格参考状态机的实现方式，而可以更为灵活。

阅读手记：

2.3.19 如何连接和驱动双向端口信号？

在连接和驱动双向端口（inout）信号时，分几种情况（主要就测试平台连接的情况而言）。如果是把两个具备 inout 端口的设计连接在一起，这种情况较易处理。如果拿到的是一个 IP 模块，该模块（具备单向端口）的目标是将来被集成到顶层（具备双向驱动端口），那么在验证模块 IP 时，需要在测试平台就这些单向驱动端口做双向驱动的"胶水逻辑"（glue logic）。这部分逻辑可以先封装在一个临时模块再与其他双向驱动模块集成，或直接将这部分逻辑在测试平台实现。

关键词：
inout，drive，multi-drive

避坑指南：
双向驱动端口的连接往往见于芯片系统测试平台。在连接时，需要考虑其他单向驱动逻辑与双向驱动端口之间的驱动关系，避免在同一时间对端口出现多个驱动源而引发多驱动问题（可能进一步导致 X 传播）。为避免发生多驱动问题，往往需要借助其他信号，或约定时序和驱动强度等来完成。

参考代码： sv_inout_multidrive_connection.sv

```
module rkv_mod1(inout io);
  logic out0;
  assign io = out0 !== 1'bz ? out0 : 1'bz;
  // behavior model
  initial begin
```

```
    forever begin
      out0 <= 1'bz;
      #($urandom_range(1, 3)*1ns);
      wait(io === 1'bz);
      out0 <= $urandom_range(0, 1);
      #($urandom_range(1, 3)*1ns);
      out0 <= 1'bz;
    end
  end
endmodule

module rkv_mod2(
  input logic in0
  ,output logic out0
  ,output logic out0_en
);
  initial begin
    forever begin
      out0_en <= 0;
      #($urandom_range(2, 4)*1ns);
      wait(in0 === 1'bz);
      out0_en <= 1'b1;
      out0 <= $urandom_range(0, 1);
      #($urandom_range(2, 4)*1ns);
      out0_en <= 1'b0;
    end
  end
endmodule

module rkv_mod3(inout io);
  logic out0_en;
  logic out0;
  rkv_mod2 m1(
    .in0(io)
    ,.out0(out0)
    ,.out0_en(out0_en)
  );
  assign io = out0_en ? out0 : 1'bz;
endmodule

module rkv_top;
  wire logic io0, io1;

  rkv_mod1 m1(io0);
  rkv_mod1 m2(io0);

  rkv_mod3 m3(io1);
  rkv_mod3 m4(io1);
endmodule
```

阅读手记:

2.4 类的使用

除了类的三要素（封装、继承、多态），在实际使用中，类的函数构建、对象复制、成员访问也是常见操作。子类句柄与父类句柄之间的转换、访问范围以及方法继承、覆盖也会成为一些用户的困惑。接下来，我们围绕着使用类的各方面来论述遇到的部分疑难点。

2.4.1 类的成员变量在声明时或在 new 函数中初始化有何区别？

从实现结果来看，这两种方式都可以对成员变量在对象创建时完成初始化。但如果以上两个动作均发生了，那么应该注意变量声明初始化执行在前（空间开辟动作），而 new 函数对成员变量初始化在后（构建函数体内部对成员变量做初始化操作）。

关键词：
variable declaration 变量声明，new 构建函数，initialization 初始化

避坑指南：
弄清楚两者的执行顺序，按照代码习惯只选择一种即可（建议使用 new() 函数做初始化赋值）。

参考代码： sv_value_init_new_order.sv

```
package rkv_pkg;
  class rkv_data;
    int data = 10;
    function new(int val = 1);
      data = val;
    endfunction
  endclass

  class rkv_trans extends rkv_data;
    int num = 5;
    int darr[];
    function new(int val = 2);
      super.new(val);
      num = 3;
      darr = new[num];
      foreach(darr[i]) darr[i] = data;
    endfunction
  endclass
endpackage

module tb;
  import rkv_pkg::*;
  rkv_trans t;
```

```
    initial begin
      t = new(3);
      $display("rkv_trans t.num = %0d", t.num);
      $display("rkv_trans t.data = %0d", t.data);
      $display("rkv_trans t.darr = %p", t.darr);
      $finish();
    end
endmodule
```

仿真结果：

```
rkv_trans t.num = 3
rkv_trans t.data = 3
rkv_trans t.darr = '{3, 3, 3}
```

阅读手记：

2.4.2 new 函数与其他函数有哪些不同？

每一个类都必须具备一个 new() 函数。用户可以不单独定义 new() 函数，但系统会在编译时为其隐式声明一个空白实现的 new() 函数。在调用 new() 函数时，其会先开辟空间，且伴随成员变量的声明和初始化，继而进入 new() 函数体执行语句。new() 函数定义时不能指定返回类型，而在调用时其返回的是该对象的句柄。

关键词：
new 函数，initialization

避坑指南：
在定义子类时，如果不实现 new() 函数，那么隐式声明的 new() 函数不带有任何参数，这可能会与父类 new() 函数的参数不一致。子类与父类的 new() 函数参数不一致本身不会有问题，但需要考虑在子类 new() 函数中如何正确地调用父类 new() 函数。一般情况下，子类应该显式地调用父类的 new 函数（以及正确地传入参数），即 super.new()，否则可能会遇到编译错误。

参考代码： sv_subclass_new_def.sv

```
package rkv_pkg;
  class packet;
    int data[];
    function new(int size);
      data = new[size];
    endfunction
```

```
    endclass
    class user_packet extends packet;
      // subclass needs to define newe() and call super.new
      function new(int size);
        super.new(size);
      endfunction
    endclass
endpackage

module tb;
  import rkv_pkg::*;
  user_packet up;
  initial begin
    up = new(3);
  end
endmodule
```

阅读手记：

2.4.3 关键词 new 的使用场景有哪些？

对于对象、旗语（semaphore）、信箱（mailbox），在创建时均需要 *new()* 函数，因此可以将旗语和信箱的实例也看成"对象"；对于动态数组，在重新开辟空间时（resize），需要 *new[]* 操作。注意，对象例化时使用的 *new()* 函数需要用小括号，而动态数组开辟空间使用的 *new[]* 需要用中括号。

关键词：
new()，new[]

避坑指南：
event 对象的创建，是不需要 *new()* 函数的，即声明 event 变量时已为其开辟了空间。

参考代码： sv_new_usage.sv

```
module tb;
  class rkv_packet;
  endclass

  initial begin
    rkv_packet pkt = new();
    int arr [] = new[3];
    mailbox #(int )bx = new(1);
    semaphore key = new(1);
```

```
          event ev;
      end
  endmodule
```

阅读手记：

2.4.4 对象占用的空间什么时候会被回收？

在创建对象时的内存空间开辟的行为，必须伴随着 *new()* 函数（构建函数）。SystemVerilog 对对象空间的管理是自动安排的，即，如果环境中没有任何一个句柄再指向该对象时，那么为其开辟的空间将会被回收，从而实现仿真时内存的动态管理。

关键词：
object　对象，handle，空间回收

避坑指南：
SystemVerilog 中的对象不需要析构函数（destructor）来释放空间。

阅读手记：

2.4.5 this 的使用场景有哪些？

this 是为当前类预定义的对象句柄，其指向当前类的域（scope）。通过使用 this，可以明确指示其所访问的目标变量或方法是该类的成员，而非来自外部的域（例如，package 中声明的变量或方法）。this 可在非静态方法、约束、内嵌约束（inline constraint）或类中定义的 covergroup 内使用。

关键词：
this，scope，member，域指向

避坑指南：
如果 this 指向的目标成员在当前类中不存在，那么会在其父类中继续逐层查找。

参考代码： sv_this_usage.sv

```
  package rkv_pkg;
    class rkv_packet;
```

```
    rand int unsigned data;
    rand int unsigned addr;
    function new();
      this.data = 0;
      this.addr = 0;
    endfunction
  endclass

  class user_packet extends rkv_packet;
    function int unsigned set_data(int unsigned data);
      this.data = data;
    endfunction
    function int unsigned get_data();
      return this.data;
    endfunction
  endclass
endpackage

module tb;
  import rkv_pkg::*;
  user_packet up;
  initial begin
    int unsigned data = 100;
    int unsigned addr = 'h10;
    up = new();
    assert(up.randomize() with {
      this.data == local::data;
      this.addr == local::addr;})
    else $error("user_packet randomization failure");
    $display("up.data = %0d, up.addr = 'h%0x", up.data, up.addr);
    $finish();
  end
endmodule
```

仿真结果：

```
up.data = 100, up.addr = 'h10
```

阅读手记：

2.4.6　对象拷贝的种类如何划分？

对象拷贝分为浅拷贝（shallow copy）和深拷贝（deep copy）。浅拷贝可直接在调用 *new()*

函数时随目标句柄完成复制（例如，h_tgt = new() h_src），该复制操作只对对象中的成员变量完成数值复制。如果被复制对象中存在句柄（指向其他对象），深拷贝可在复制时克隆被复制对象成员句柄所指向的对象。

关键词：
shallow copy　浅拷贝，deep copy　深拷贝

避坑指南：
深拷贝需要用户自己实现其方法，并考虑递归复用子类和父类的深拷贝方法。

参考代码： sv_object_copy_shallow_deep.sv

```
package rkv_pkg;
  class transaction;
    bit[31:0] addr;
    bit[31:0] data;
    function new(bit[31:0] addr, bit[31:0] data);
      this.addr = addr;
      this.data = data;
    endfunction
    function void print(string name = "transaction");
      $display("%s addr = %0x, data = %0x", name, addr, data);
    endfunction
    function transaction copy();
      transaction t = new this;
      return t;
    endfunction
  endclass

  class packet;
    transaction tr1, tr2;
    function new(bit[31:0] addr, bit[31:0] data);
      tr1 = new(addr, data);
      tr2 = new tr1;
    endfunction
    function void print(string name = "packet");
      $display("%s sub children content is as follow:", name);
      tr1.print("tr1");
      tr2.print("tr2");
    endfunction
    function packet copy();
      packet p = new(0, 0);
      p.tr1 = tr1.copy();
      p.tr2 = tr2.copy();
      return p;
    endfunction
  endclass
endpackage
```

```
module tb;
  import rkv_pkg::*;
  transaction tr1, tr2;
  packet p1, p2;
  initial begin
    tr1 = new('h404, 'h302);
    tr1.print("tr1");
    tr2 = new tr1;
    tr2.print("tr2");
    p1 = new('h40, 'h20);
    p1.print("p1");
    p2 = p1.copy();
    p2.print("p2");
  end
endmodule
```

仿真结果：

```
tr1 addr = 404, data = 302
tr2 addr = 404, data = 302
p1 sub children content is as follow:
tr1 addr = 40, data = 20
tr2 addr = 40, data = 20
p2 sub children content is as follow:
tr1 addr = 40, data = 20
tr2 addr = 40, data = 20
```

阅读手记：

2.4.7 this 和 super 有什么联系和差别？

this 所查找的范围为当前类及其所继承的父类（及以上层次），即该类声明或继承的所有成员，并且按照变量就近索引的原则（自当前子类往上逐级查找）；super 查找范围限定于该子类的父类（及以上层次），适用于访问某些被子类覆盖的成员（变量或方法）。这两种关键词的查找方位均限定于该类的继承层次，与外部的、引入的变量或方法无关。

关键词：

this，super

避坑指南：

不能使用 super.super 的形式去定向访问更上层的成员。

参考代码：sv_this_diff_super.sv

```
module tb;
  class c0;
    string type_self = "c0";
    string type_shared = "c1";
  endclass
  class c1 extends c0;
    string type_self = "c1";
  endclass
  class c2 extends c1;
    string type_self = "c2";
    function new();
      this.type_shared = "c2";
      $display("type_shared = %s", super.type_shared);
      $display("this.type_self = %s", this.type_self);
      $display("super.type_self = %s", super.type_self);
    endfunction
  endclass

  initial begin
    c2 c2_inst = new();
    $finish();
  end
endmodule
```

仿真结果：

```
type_shared = c2
this.type_self = c2
super.type_self = c1
```

阅读手记：

2.4.8 子类成员如何覆盖父类成员？

子类成员变量可声明与父类成员同名的变量（类型可不同），去覆盖父类成员变量。子类成员方法也可以声明同名方法去覆盖父类成员方法。覆盖成员方法时，如果不添加 virtual 描述符，那么方法的参数类型和数目、有无返回值等形式可不同。在这一情况下，父类和子类的成员方法不通过多态（polymorphism）进行动态查找（dynamic lookup）。

关键词：

member variable　成员变量，member method　成员方法，override　覆盖，polymorphism，dynamic lookup

避坑指南：

建议保持子类成员方法与父类成员方法在形式上的"三者合一"，即它们的方法名、参数类型和数目、有无返回值这三者保持一致，从而为后续的多态绑定（即虚方法的用途）做准备。

参考代码：sv_class_child_override_parent.sv

```
module tb;
  class parent;
    bit [7:0] vec1 = 8'h7;
    virtual function bit[7:0] get_val();
      return vec1;
    endfunction
  endclass

  class child extends parent;
    bit [15:0] vec1 = 16'hff;
    // SAME 1) method name 2) args name & num 3) return type
    // polymorphism with virtual method declaration
    function bit[7:0] get_val();
      return vec1;
    endfunction
  endclass

  initial begin
    bit [7:0] val;
    parent p1 = new();
    child c1 = new();
    $display("p1 obj get_val() = 'h%0x", p1.get_val());
    $display("c1 obj get_val() = 'h%0x", c1.get_val());
    p1 = c1;
    $display("c1 obj get_val() = 'h%0x", p1.get_val());
    $finish();
  end
endmodule
```

仿真结果：

```
p1 obj get_val() = 'h7
c1 obj get_val() = 'hff
c1 obj get_val() = 'hff
```

阅读手记：

2.4.9 子类句柄和父类句柄的类型如何转换？

子类句柄可以采用'='直接赋值给父类句柄，在赋值时发生了类型的隐式转换；父类句柄如果要转换为子类句柄，必须使用$cast(TGT, SRC)类型转换函数，且转换成功的前提（$cast()返回'1'）是父类句柄指向的是子类对象。

关键词：
handle type casting　句柄类型转换，subclass　子类，superclass（parent class）　父类

避坑指南：
在父类句柄（指向子类对象 A）通过$cast()转换为子类句柄时，无论子类句柄之前是否已经指向另一个子类对象 B，在转换成功后，子类句柄和父类句柄都指向子类对象 A。

参考代码： sv_class_child_parent_handle_cast.sv

```
package rkv_pkg;
  class rkv_packet;
    int obj_id;
    static int _id = 0;
    function new(string tname = "rkv_packet");
      if(tname == "rkv_packet") begin
        obj_id = _id;
        _id++;
      end
    endfunction
    virtual function int get_id();
      return obj_id;
    endfunction
    virtual function string get_name();
      return "rkv_packet";
    endfunction
  endclass

  class user_packet extends rkv_packet;
    static int _id = 0;
    function new(string tname = "user_packet");
      super.new(tname);
      if(tname == "user_packet") begin
        obj_id = _id;
        _id++;
      end
    endfunction
    virtual function string get_name();
      return "user_packet";
    endfunction
  endclass
endpackage
```

```
module tb;
  import rkv_pkg::*;
  parameter RP_NUM = 3;
  parameter UP_NUM = 5;
  rkv_packet rp[RP_NUM], trp;
  user_packet up[UP_NUM], tup;
  initial begin
    foreach(rp[i]) rp[i] = new();
    foreach(up[i]) up[i] = new();
    tup = up[1];
    $display("tup -> up[1] with name [%s] id [%0d]",
             tup.get_name(), tup.get_id());
    trp = up[2];
    $display("trp -> up[2] with name [%s] id [%0d]",
             trp.get_name(), trp.get_id());
    assert($cast(tup, trp))
    else $error("type casting failure!");
    $display("tup -> up[2] with name [%s] id [%0d]",
             tup.get_name(), tup.get_id());
    $finish();
  end
endmodule
```

仿真结果：

```
tup -> up[1] with name [user_packet] id [1]
trp -> up[2] with name [user_packet] id [2]
tup -> up[2] with name [user_packet] id [2]
```

阅读手记：

2.4.10 为什么父类句柄经常需要转换为子类句柄？

在不考虑多态（虚方法）的情况下，父类句柄在指向一个子类对象时，它能访问的只有父类成员，而如果要访问子类成员，那么它就应该转换为子类句柄。

关键词：

handle type casting　句柄转换，accessible scope　访问范围

避坑指南：

父类句柄要访问子类成员方法，可以通过虚方法来完成动态查找（dynamic lookup），不一定要转为子类句柄。而父类句柄要访问子类成员变量，则除了转为子类句柄，没有其他方法。

参考代码：sv_class_parent_handle_access_child_member.sv

```
package rkv_pkg;
  class rkv_packet;
    int data = 'h10;
    int addr = 'h20;
  endclass

  class user_packet extends rkv_packet;
    int ufield = 'h30;
  endclass
endpackage

module tb;
  import rkv_pkg::*;
  rkv_packet trp;
  user_packet up, tup;
  initial begin
    up = new();
    trp = up;
    $display("trp -> up could access data 'h%0x, addr 'h%0x",
             trp.data, trp.addr);
    void'($cast(tup, trp));
    $display("tup -> up could access data 'h%0x, addr 'h%0x, ufield 'h%0x",
             tup.data, tup.addr, tup.ufield);
  end
endmodule
```

仿真结果：

```
trp -> up could access data 'h10, addr 'h20
tup -> up could access data 'h10, addr 'h20, ufield 'h30
```

阅读手记：

2.4.11 虚方法声明和不声明的区别是什么？

虚方法（virtual method）的声明就是为了便于父类句柄在访问子类对象的成员方法时，不需要通过句柄转换即可实现动态查找（dynamic lookup）。如果没有在成员方法前添加 virtual 描述符，那么该方法的动态查找将无法实现。

关键词：
virtual

避坑指南：

除了 *new()* 函数不能添加 virtual，父类的其余方法均可添加 virtual 描述符。类的成员变量不能添加 virtual 描述符。

参考代码： sv_class_virtual_dynamic_binding.sv

```
package rkv_pkg;
  class rkv_packet;
    virtual function string get_name();
      return "rkv_packet";
    endfunction
  endclass

  class user_packet extends rkv_packet;
    function string get_name();
      return "user_packet";
    endfunction
  endclass
endpackage

module tb;
  import rkv_pkg::*;
  rkv_packet rp;
  user_packet up;
  initial begin
    up = new();
    rp = up;
    $display("up -> user_packet instance and got name [%s]",
             up.get_name());
    $display("rp -> user_packet instance and got name [%s]",
             rp.get_name());
    $finish();
  end
endmodule
```

仿真结果：

```
up -> user_packet instance and got name [user_packet]
rp -> user_packet instance and got name [user_packet]
```

阅读手记：

2.4.12 虚方法的描述符 virtual 应该在哪里声明？

如果需要类的成员方法完成动态查找（绝大多数情况下是这么建议的），那么应该在根父类（root superclass）对该方法添加 virtual 描述符，而至于子类及后续子类的同名方法，则不要求添加 virtual 描述符，并将其同样视为虚方法属性。

关键词：

virtual，superclass，subclass

避坑指南：

在实现子类的过程中，如果要添加新的虚方法，那么应将该虚方法的 virtual 声明添加到父类中。

参考代码： sv_class_virtual_declaration_hier.sv

```
package rkv_pkg;
  class rkv_packet;
    rand int data[];
    constraint cstr{data.size() inside {[2:4]};}
    virtual function string get_name();
      return "rkv_packet";
    endfunction
    function extends_data();
      print_data_size();
      // extend size + 1
      data = new[data.size()+1] (data);
      $display("%s::extends_data() method-1 is called", get_name());
      print_data_size();
    endfunction
    function print_data_size();
      $display("%s::data size is %0d", get_name(), data.size());
    endfunction
  endclass

  class user_packet extends rkv_packet;
    function string get_name();
      return "user_packet";
    endfunction
    virtual function extends_data();
      print_data_size();
      // extend size + 2
      data = new[data.size()+2] (data);
      $display("%s::extends_data() method-2 is called", get_name());
      print_data_size();
    endfunction
  endclass

  class test_packet extends user_packet;
    function string get_name();
      return "test_packet";
    endfunction
```

```
    function extends_data();
      // extend size + 3
      super.extends_data();
      data = new[data.size()+1] (data);
      $display("%s::extends_data() method-3 is called", get_name());
      print_data_size();
    endfunction
  endclass
endpackage

module tb;
  import rkv_pkg::*;
  rkv_packet trp;
  user_packet up, tup;
  test_packet tp;
  initial begin
    up = new(); void'(up.randomize());
    tp = new(); void'(tp.randomize());
    trp = up;
    trp.extends_data();
    trp = tp;
    trp.extends_data();
    tup = tp;
    tup.extends_data();
    $finish();
  end
endmodule
```

仿真结果：

```
user_packet::data size is 2
user_packet::extends_data() method-1 is called
user_packet::data size is 3
test_packet::data size is 2
test_packet::extends_data() method-1 is called
test_packet::data size is 3
test_packet::data size is 3
test_packet::extends_data() method-2 is called
test_packet::data size is 5
test_packet::extends_data() method-3 is called
test_packet::data size is 6
```

阅读手记：

2.4.13 句柄一旦不指向对象，该对象就被回收了吗？

按照 SystemVerilog 的标准，一旦没有任何句柄指向目标对象，那么该对象占用的空间将被回收。在实际使用中，这个回收过程是由仿真器来安排控制的。很多时候，仿真器会将本应回收的对象空间继续保留一段时间，而在仿真资源紧张时，则将这些本应回收的对象真正回收、释放空间。

关键词：
object reclaim　对象回收，automatic memory management

避坑指南：
任何对象，一旦句柄与它失去了联系，那么就再也找不回来。

阅读手记：

2.4.14 什么时候该使用 local 和 protected？

local 与 protected 的封装属性，都可以隔离对外部对类的成员的访问。对于验证环境，反倒一般不建议添加 local 和 protected，因为一般验证环境的层次没有太深、太复杂。在开发验证环境时，一开始可以先放开访问权限，这有利于早期的环境稳定，待环境稳定后，可以考虑对哪些内部方法加以 local、protected 的描述限定。这部分的建议与传统的软件开发的观点有所不同。

关键词：
local，protected，verification environment，stable，accessible

避坑指南：
如果你的 VIP 验证组件或者模块环境，接下来会被别人"复用"，那么你应该在正式发布时考虑添加 local、protected 等访问限定描述符。这样可以对他人使用或集成你的验证组件加以限定，方便验证组件的后期维护。

阅读手记：

2.4.15 class 和 virtual class 的区别是什么？

用户自定义的类一般都采用 class 而不采用 virtual class。我们的验证项目框架的尺寸一般不会像大型软件项目那样，这就使得用户在构建类的层次时并不太在意 virtual class 需要在什么时间使用。

在构建一个抽象类（内部成员不完整，只声明方法但未对其实现）时，我们希望子类能够实现这些方法，因此这个抽象类（virtual class）就提供了一个早期的原型（prototype）。并且由于它的不完整，它本身也不能做例化。而继承于抽象类的子类，必须实现其父类声明的方法原型（pure virtual method）。同时，在必要的情况下，它的子类或后续的继承子类，还可以在继承链条中的某一环中作为抽象类，以此规定后续子类应该实现的方法。所以，一般而言，如果是定义抽象类，那么往往伴随着 pure virtual method。

关键词：

class，virtual class，pure virtual method

避坑指南：

pure virtual method 只能在 virtual class 中声明；pure 关键词只能对 virtual method 进行声明，而不能对 virtual class 进行声明；virtual class 的子类必须实现其父类的 pure virtual method；virtual class 不能例化。

参考代码：

```
module tb;
  virtual class abstract_packet;
    pure virtual function void content();
  endclass
  virtual class base_packet extends abstract_packet;
    function void content();
    endfunction
    pure virtual function void format();
  endclass
  class packet extends base_packet;
    function void format();
    endfunction
  endclass
  initial begin
    // ERROR abstract class cannot be instantiated
    // abstract_packet ap = new();
    // base_packet bp = new();
    packet p = new();
  end
endmodule
```

阅读手记：

2.4.16 virtual 修饰符在哪些场景中会用到？

virtual 可以用来在类中声明接口句柄，即 virtual interface，接口句柄在使用（索引目标接口内的变量）时需要确保它不为空（null），否则将出现运行时错误。virtual 也可以在类中声明虚方法（virtual method），虚方法可以实现方法调用时的动态查找（dynamic lookup），即父类句柄可以调用子类的同名方法。virtual 还可以用来修饰类，表示一个抽象类，该类本身不能例化，但可以通过继承它来实现或添加方法。此外，pure virtual 可以在 virtual class 中用来修饰方法，纯虚方法只需要声明不需要实现（在子类中需要实现）。

关键词：

virtual interface，virtual method，virtual class，pure virtual method

避坑指南：

virtual interface 的声明通常出现在类中，在赋值后用来在 TB 和 DUT 之间进行数据驱动和监测。它也可以在 module 中声明，用 virtual interface 数组来管理多个接口实例，实现批量化处理。

参考代码： sv_interface_virtual_reference.sv

```
interface intf;
  logic a;
  logic b;
endinterface

class driver;
  virtual intf vif;
endclass

module tb;
  intf ifi0();
  intf ifi1();
  intf ifis[2] ();
  virtual intf vifs[4];
  driver drv;
  initial begin
    vifs = '{ifi0, ifi1, ifis[0], ifis[1]};
    vifs[0].a = 1'b1;
    drv = new();
    drv.vif = ifi1;
    drv.vif.b = 1'b0;
    $finish();
  end
endmodule
```

阅读手记：

2.4.17　子类能够使用与父类相同名称但不同参数的方法吗？

在 SystemVerilog 的标准中尽管没有明确规定子类方法与父类方法保持一致，但为了能够实现动态查找（dynamic lookup），我们还是建议保持"三个一致"，即子类方法与父类方法的名称、参数和返回值类型均保持一致。如果出现不一致的情况，那么父类无法对同名方法声明 virtual，也即无法实现虚方法的动态查找。同时，在类中不能声明多个同名的方法（例如，采用不同参数或不同返回值的形式）。

关键词：
inheritance　继承，consistence，virtual method　虚方法，dynamic lookup　动态查找

避坑指南：

个别情况下，如果不需要实现虚方法的动态查找，那么子类和父类的同名方法可以不遵循"三个一致"。在这种情况下，子类方法仍然可以通过 *super.METHOD()* 来调用父类方法，实现对父类方法的继承。不过，在大多数情况下，我们并不建议这样做。考虑到 UVM 环境中的应用场景，更多的是在父类中声明 *virtual METHOD*，而在子类中保持与父类同名方法的"三个一致"，以便于后期对同名方法的调用以及类型覆盖等。

参考代码： sv_class_method_inheritance.sv

```systemverilog
module tb;
  class pkt;
    protected byte unsigned _data;
    function new(int unsigned ini = 'h33);
      _data = ini;
    endfunction
    function byte unsigned get_data();
      return _data;
    endfunction
  endclass
  class enc_pkt extends pkt;
    function shortint unsigned get_data(bit enc = 0);
      get_data = enc ? (_data << 1) : super.get_data();
    endfunction
  endclass

  initial begin
    enc_pkt ept = new();
    pkt pt = ept;
    $display("ept member _data is 'h%0x via ept.get_data()",
             ept.get_data(1));
    $display("ept member _data is 'h%0x via pt.get_data()",
             pt.get_data());
  end
endmodule
```

仿真结果：

```
ept member _data is 'h66 via ept.get_data()
ept member _data is 'h33 via pt.get_data()
```

阅读手记：

2.4.18 使用参数类或者接口时要注意哪些？

在使用参数类或者接口时，需要考虑，如果参数是 type T 类型（即类型可通过参数转变），那么在接下来的声明句柄、例化、配置等阶段，都需要严格指定 type T 类型，以保证源端（source）和目标端（target）通过 type T 指定的类型是一致的，否则可能会在编译或运行时发生错误。

关键词：

parameterized，type T，class，interface，config_db

避坑指南：

如果参数是 type T，那么在声明虚接口、句柄时，一定要给定相同的类型参数，这样才能满足编译需要。如果参数是可以设置数值的（如 int 类型），那么要看参数值的给定是否影响目标的数据结构、逻辑结构等。

不同的仿真器（VCS、Questasim）对参数类的句柄赋值都要求严格，即对参数类型，无论是普通类型（例如 int）还是 type T 类型，都要求左右两侧句柄保持严格一致。但是，对于像虚接口的声明和赋值等，则会出现差异。总体而言，我们建议对于参数类或参数接口，在进行句柄赋值或虚接口赋值时，应该保持左右两侧指定的参数一致，避免编译或运行时发生错误。

参考代码： sv_parameterized_type_class_interface.sv

```
interface rkv_intf1 #(type T = int);
  T data[];
endinterface
interface rkv_intf2 #(int N = 1);
  int data[N]; // N would effect data structure
  int val = N; // N would not effect data structure
endinterface

module tb;
  class rkv_packet1 #(type T = int);
    T data[];
  endclass
```

```
class rkv_packet2 #(int N = 1);
  int data[N]; // N would effect data structure
  int val = N; // N would not effect data structure
endclass

rkv_intf1                intf0();
rkv_intf1 #(string)      intf1();
rkv_intf2                intf2();
rkv_intf2 #(3)           intf3();

initial begin
  virtual rkv_intf1              v_intf0;
  virtual rkv_intf1 #(string)    v_intf1;
  virtual rkv_intf2              v_intf2;
  virtual rkv_intf2 #(3)         v_intf3;
  v_intf0 = intf0;
  v_intf1 = intf1;
  v_intf2 = intf2;
  v_intf3 = intf3;
end

initial begin
  rkv_packet1              pkt0;
  rkv_packet1 #(string)    pkt1;
  rkv_packet2              pkt2;
  rkv_packet2 #(3)         pkt3;
  rkv_packet1              h_pkt0;
  rkv_packet1 #(string)    h_pkt1;
  rkv_packet2              h_pkt2;
  rkv_packet2 #(3)         h_pkt3;
  pkt0 = new();
  pkt1 = new();
  pkt2 = new();
  pkt3 = new();
  h_pkt0 = pkt0;
  h_pkt1 = pkt1;
  h_pkt2 = pkt2;
  h_pkt3 = pkt3;
end
endmodule
```

阅读手记:

2.5 随机约束使用

要随机产生一个数据或对若干变量随机化,又或对某一个对象完成随机化,实现起来都有对应的办法。而随机属性 rand 一旦与句柄、数组、结构体关联,其表现仍然有必要进行探讨。与随机有关的约束,如果发生了约束的冲突、继承,其又会如何表现?这也需要进一步梳理。接下来,我们将就随机约束的使用问题论述若干疑难点。

2.5.1 rand 描述符可用于哪些变量类型?

首先,rand 描述符只允许在 class 域中使用。所有向量(logic、bit、int、integer、byte 等)都可以用 rand 描述,由向量构成的组合或非组合数据结构(各类数组、结构体)也可用 rand 描述,句柄也可用 rand 描述。

关键词:
rand,vector,array,struct,handle

避坑指南:
string、real 类型不应该用 rand 修饰,logic 中的 X、Z 值无法随机化产生。

参考代码: sv_class_randomize.sv

```
package rkv_pkg;
  class packet;
    rand bit [31:0] src;
    rand logic [31:0] dst;
    rand int unsigned data [8];
    randc byte unsigned kind;
    byte unsigned addr;
    constraint cstr {
      src inside {[10:15]};
    }
  endclass
endpackage

module tb;
  import rkv_pkg::*;
  packet p;
  initial begin
    p = new();
    p.cstr.constraint_mode(0);
    assert(p.randomize(addr) with {
           addr[1:0] == 0;
           addr inside{['h4:'h10]};})
    else
```

```
        $error("packet randomization failure!");
      $finish();
    end
  endmodule
```

阅读手记：

2.5.2 数组使用 rand 声明会发生什么？

可随机化的数组（动态数组、队列、关联数组）在随机化时，其大小和元素值均会发生变化。数组在随机化时，会先后对数组大小和数组元素进行随机化（独立的两个步骤）。即随机时，先利用原有数组内容对新数组进行初始化（扩展或截取），接下再对每个元素值进行随机化。

关键词：

rand，array 数组，size，element

避坑指南：

对可随机化数组的约束应限定其大小，并在需要时限定其每个元素的取值。在特定情况下，可通过 rand_mode 禁止数组中某些元素的随机属性。

参考代码： sv_rand_array.sv

```
module tb;
  class packet;
    rand bit[3:0] dyn_arr[];
    rand bit[3:0] que_arr[$];
    constraint cstr {
      dyn_arr.size inside {[2:5]};
      que_arr.size == dyn_arr.size();
    }
  endclass

  initial begin
    packet p = new();
    p.randomize() ;
    $display("p content = %p", p);
    p.dyn_arr.rand_mode(0);
    p.randomize() ;
    $display("p content = %p", p);
    p.rand_mode(1);
    p.randomize() with {dyn_arr.size() == 3;};
```

```
      $display("p content = %p", p);
    end
endmodule
```

仿真结果：

```
p content = '{dyn_arr:'{'h3, 'hf, 'h1, 'ha, 'h4} ,
             que_arr:'{'h3, 'h2, 'h9, 'he, 'h0} }
p content = '{dyn_arr:'{'h3, 'hf, 'h1, 'ha, 'h4} ,
             que_arr:'{'h6, 'h7, 'h6, 'h1, 'h3} }
p content = '{dyn_arr:'{'h1, 'he, 'h3} ,
             que_arr:'{'hc, 'h9, 'h2} }
```

阅读手记：

▶▶ 2.5.3 句柄使用 rand 声明会发生什么？

如果将句柄声明为 rand，那么它所在的对象在随机化时（外部通过 *randomize()* 函数调用），该句柄所指向的对象也会被触发随机化，继而可通过随机化句柄的方式完成逐级递进的随机化自动调用。

关键词：
rand，randomize，handle　句柄

避坑指南：
请在调用上层对象的随机化函数时，确保其成员 rand 句柄是非悬空状态（指向具体对象，不为 null），否则该次随机化并不会为 rand 句柄创建对象并随机化，其值将依然为 null。

参考代码： sv_rand_handle.sv

```
module tb;
  class rkv_packet;
    rand bit[3:0] dyn_arr[];
    rand bit[3:0] que_arr[$];
    constraint c1 {
      dyn_arr.size inside {[2:5]};
      que_arr.size == dyn_arr.size();
    }
  endclass

  class rkv_trans;
    rand rkv_packet pkt;
    function new();
```

```
    pkt = new();
  endfunction
endclass

initial begin
  rkv_trans tr = new();
  if(!tr.randomize()) $error("randomization failure!");
  $display("tr content = %p", tr);
  $display("tr.pkt content = %p", tr.pkt);
end
endmodule
```

仿真结果：

```
tr content = '{pkt:{ ref to class rkv_packet}}
tr.pkt content = '{dyn_arr:'{'h3, 'hf, 'h1, 'ha, 'h4} ,
                   que_arr:'{'h3, 'h2, 'h9, 'he, 'h0} }
```

阅读手记：

2.5.4 rand 和 randc 的区别在哪里？

rand 变量在每一次随机化时，与之前的随机化取值都没有关联；randc 则会排除之前随机化的数值，直到所有可能的数值都随机化遍历，才进行下一轮的随机化。如果用抽扑克牌来打比方的话，使用 rand 每次随机化时都会将所有的牌重新洗好再随机抽出，而使用 randc 则会在每次抽出牌后继续抽出接下来的牌，直到牌全部抽完才重新洗牌。

关键词：

rand，randc

避坑指南：

使用 randc 做随机化时，应保持使用同一个对象的句柄，调用其 randomize()函数。

参考代码： sv_rand_diff_randc.sv

```
module tb;
  typedef enum {SINGLE, BURST, INCR, WRAP, OUTORDER} command_t;
  class packet;
    rand command_t cmd_r;
    bit cmd_r_arr[command_t];
    randc command_t cmd_rc;
    bit cmd_rc_arr[command_t];
    int rtimes;
```

```
      function void post_randomize();
        cmd_r_arr[cmd_r] = 1;
        cmd_rc_arr[cmd_rc] = 1;
        rtimes++;
      endfunction
      typedef bit cmd_arr_t [command_t];
      function bit is_command_complete(ref cmd_arr_t arr);
        command_t rq [$] = {SINGLE, BURST, INCR, WRAP, OUTORDER};
        command_t q[$];
        foreach(arr[cmd]) q.push_back(cmd);
        return (rq == q);
      endfunction
      function void rand_command_complete(ref cmd_arr_t arr);
        forever begin
          void'(this.randomize());
          if(is_command_complete(arr)) begin
            $display("rand(c) array completed with %0d times randomization",
                     rtimes);
            return;
          end
        end
      endfunction
    endclass

    initial begin
      packet p = new();
      p.rand_command_complete(p.cmd_rc_arr);
      p.rand_command_complete(p.cmd_r_arr);
    end
  endmodule
```

仿真结果：

```
rand(c) array completed with 5 times randomization
rand(c) array completed with 12 times randomization
```

阅读手记：

2.5.5 内嵌约束中的 local:: 表示什么？

在使用内嵌约束（inline constraint）即 *randomize() with {CONSTRAINT}* 时，约束体中变量名的查找顺序默认是从被随机化对象开始查找的。但如果调用 *randomize()* 时所处的局部域（local scope）中也有同名变量，那么就需要使用 local:: 来显式声明该变量来源于外部，而非被随机化的对象（在约束中也可用 this、super 来索引这些变量）。

关键词：

Inline constraint，randomize，local::，local scope，this

避坑指南：

local::只表示"域"，并不指代某个句柄，所以也可用 local::this 表示调用 randomize()函数时所在对象的句柄。

参考代码： sv_inline_constraint_local_scope.sv

```
module tb;
  class c1;
    rand int m_type = 0;
    rand int m_id = 0;
    constraint cstr {soft m_type == 1; soft m_id == 1;}
  endclass

  class c2 extends c1;
    c1 c1_inst;
    function new();
      c1_inst = new();
      m_type = 2;
      m_id = 2;
      c1_inst.randomize with {
        m_type == local::this.m_type;
        m_id == local::this.m_id;
      };
    endfunction
  endclass

  initial begin
    c2 c2_inst = new();
    $display("c2 body : %p \n c2_inst.c1_inst body : %p",
             c2_inst, c2_inst.c1_inst);
    $finish();
  end
endmodule
```

仿真结果：

```
c2 body : '{m_type:2, m_id:2, c1_inst:{ ref to class c1}}
c2_inst.c1_inst body : '{m_type:2, m_id:2}
```

阅读手记：

2.5.6 是否可以利用动态数组对变量值的范围进行约束？

当然可以，而且这么做的话，你还可以在每次随机化之前，通过对动态数组内容做修改，影响每次随机时的约束范围。只不过，由于仿真器的编译限制，应该将整个动态数组作为你的变量值的约束范围，而不是进行片段选择（slice select）。例如，constraint {a inside {dynarr[M:N]};}，正确的使用方式应该是 constraint {a inside dynarr;}。

关键词：

constraint 约束，dynamic array 动态数组

避坑指南：

如果确实要对动态数组进行片选，那么可以另外声明一个动态数组，如 dynsub，并将 dynarr 数组的片段完整赋值给 dynsub，即 dynsub = dynarr[M:N]。然后将 dynsub 代入到约束中，即 constraint {a inside dynsub;}，即可达到目的。

参考代码： sv_constraint_with_dyn_array.sv

```
module tb;
  class constr_arr;
    int dynarr [], dynsub[];
    rand int a, b;
    constraint cstr {
      a inside {dynsub};
      b inside dynarr;
    }
    function void pre_randomize();
      dynsub = dynarr[0:2];
    endfunction
  endclass

  initial begin
    constr_arr c1 = new();
    // pre-define elements for constraint solution
    c1.dynarr = '{10,20,30,40,50,60}; // 6 elements
    c1.randomize();
    $display("c1.a= %0d", c1.a);
    $display("c1.b= %0d", c1.b);
  end
endmodule
```

仿真结果：

```
c1.a= 10
c1.b= 40
```

阅读手记：

2.5.7 多个软约束在随机化时有冲突是否可以解决？

首先可以肯定的是，如果一些软约束（soft constraint）之间有冲突，那么在随机化时是可以成功的。随机化解决冲突，是通过后置约束覆盖前置约束、子类约束覆盖父类约束，以及外部约束（即内嵌约束，inline constraint）覆盖类内部约束的方式来完成的。

例如，下面代码中的 cstr2 与 cstr1 有冲突，那么 cstr2 覆盖 cstr1。在 p 随机化时，外部约束与 cstr2 共同得出变量只可能取值 *v = 40*。如果此时外部约束与 *packet::cstr2* 有冲突，那么外部约束将覆盖类的内部约束，并且生效。

关键词：
soft，constraint，conflict

避坑指南：
如果采取强约束（默认约束类型）覆盖软约束，或者多个软约束按照层次关系来解决约束冲突，就可以在不产生随机化失败的同时解决约束冲突问题。在类的层次化继承中，或 sequence 的多个层次包裹的约束块中，都可以使用多个软约束来解决可能的约束冲突问题。

参考代码： sv_constraint_soft_solution.sv

```
module tb;
  class packet;
    rand int v;
    constraint cstr1 {
      soft v inside {[10:20]};
    };
    constraint cstr2 {
      soft v inside {[30:40]};
    };
  endclass
  initial begin
    packet p = new();
    if(!p.randomize() with {soft v inside {[40:50]};})
      $error("randomization failure!");
    else
      $display("v = %0d", p.v);
  end
endmodule
```

仿真结果：

```
v = 40
```

阅读手记：

2.5.8 结构体是否可对其成员使用 rand 描述符？

在对 struct 类型进行定义时，需要注意它是组合型（packed）还是非组合型（unpacked，默认类型）。如果是默认的非组合型，那么可以对其内部的成员根据随机化的需要指定添加 rand 描述符。如果是在定义时对结构体类型添加了 packed 关键字，那么将无法对其内部成员单独添加 rand 描述符。

此外，在类中声明结构体成员时，为了接下来的随机化，也需要对该成员添加 rand 修饰符。综上，类中的结构体变量为非组合型时，需要对该结构体变量及其内部成员分别添加 rand 修饰符以备随机化；结构体变量为组合型时，仅需要对该结构体变量添加 rand 修饰符（且不再支持对其内部成员指定添加 rand 修饰符）。

关键词：
struct，rand，packed，unpacked

避坑指南：
默认情况下的结构体类型（非组合型）在定义时，注意要对其内部成员同样添加 rand 描述符。

参考代码： sv_rand_struct.sv

```
module tb;
  typedef struct packed {
    bit [31:0] addr;
    bit [ 7:0] opcode;
    bit [ 3:0] [31:0] data ;
  } packeta_t;

  typedef struct {
    rand bit [31:0] addr;
    rand bit [ 7:0] opcode;
        bit [31:0] data [4];
  } packetb_t;

  class packet;
    rand packeta_t pkta;
```

```
    rand packetb_t pktb;
  endclass

  initial begin
    packet pkt = new();
    assert(pkt.randomize())
    else $error("randomization failure");
    $finish();
  end
endmodule
```

阅读手记：

2.5.9 如何随机化对象的多个成员且使每次数据不重复？

如果要使对象中某个变量在若干次随机中产生不同的数据，那么可以使用 randc 修饰符来实现。但如果要使对象中的多个随机属性成员在每次随机后的内容不重复，那么就无法通过给这些随机属性成员添加 randc 修饰符来实现。

令人可惜的一点是，SystemVerilog 语法并没有就这一点给出说明，所以用户需要自己实现。此外，目前已知的如 Mentor inFact、Cadence PerSpec 这类 PSS 工具，可以就覆盖率目标产生所有可能的组合，在路科的 V3 验证课程中，我们可以学到这两个工具是如何产生不同数据组合的。

关键词：

rand，randc，unique

避坑指南：

如果要随机产生的数据较多，且随机空间也较大，那么要求后续每次随机化生成的数据与之前都不重复的随机约束求解的时间将持续增加。为满足这一条件，也要求持续保留先前已生成的数据内容，这将在内存和求解效率方面带来挑战。

参考代码： sv_rand_member_unique.sv

```
module tb;
  typedef struct packed {
    bit[31:0] addr;
    bit[ 1:0] opcode;
    bit[ 7:0] size;
  } packet_info_t;
  class packet;
    rand packet_info_t info;
```

```
      rand bit[31:0] data [];
      static packet_info_t info_q[$];
      constraint cstr {
        info.addr [1:0] == 0;
        info.addr inside {['h00:'h1F]};
        info.size inside {1, 4, 8, 16, 32, 64};
        data.size() == info.size;
        unique {info_q, info};
        // same effect as 'unique' but lower solution performance
        // foreach(info_q[i]) info != info_q[i];

      }
      function void post_randomize();
        info_q.push_back(info);
      endfunction
    endclass

    initial begin
      int rt = 0;
      packet p = new();
      forever begin
        if(p.randomize())
          $display("p randomization success at %0d times", ++rt);
        else begin
          $display("p randomization failure at %0d times", ++rt);
          break;
        end
      end
      $finish();
    end
  endmodule
```

阅读手记：

2.5.10 子类会继承还是覆盖父类的约束？

子类会继承父类所有的约束，但如果子类与父类具有同名的约束，那么子类的约束将覆盖父类的约束（父类的约束在随机中将被覆盖，不再参与到随机求解过程中）。同时，在随机化过程中，无论我们使用的是父类句柄还是子类句柄，都不会影响子类对象的随机化结果。

关键词：
constraint，inheritance，override，randomize

避坑指南:
为了避免子类因与父类采取同名的约束而覆盖父类的约束,可以采用一定的约束命名规则,例如,添加与子类名相同的前缀或后缀。

参考代码: sv_class_constraint_override.sv

```
module tb;
  class base_c;
    rand byte unsigned i;
    constraint c {i inside {[10:12]};}
    virtual function void print();
    endfunction
  endclass

  class der_c extends base_c;
    rand byte unsigned i;
    constraint c {i inside {[20:22]};}
    function void print();
      $display("der_c::i = %0d, base_c::i = %0d", i, super.i);
    endfunction
  endclass

  initial begin
    der_c inst = new();
    base_c bch = inst;
    if(inst.randomize())
      inst.print();
    if(bch.randomize())
      bch.print();
  end
endmodule
```

仿真结果:

```
der_c::i = 20, base_c::i = 163
der_c::i = 20, base_c::i = 214
```

参考代码: sv_class_constraint_inheritance.sv

```
module tb;
  class packet;
    rand bit [7:0] data;
    constraint cstr1 {
      data inside {[0:31]};
    }
  endclass

  class user_packet extends packet;
    constraint cstr2 {
```

```
      data inside {[31:63]};
    }
  endclass

  initial begin
    user_packet up = new();
    repeat(3) begin
      if(!up.randomize())
        $error("up randomization failure");
      else
        $display("up.data is %0d", up.data);
    end
  end
endmodule
```

仿真结果：

```
up.data is 31
up.data is 31
up.data is 31
```

阅读手记：

2.6 覆盖率应用

定义一个覆盖组，再对其覆盖点和仓进行完善，这是实现功能覆盖率的方法。而在何处定义覆盖组、如何定义覆盖组的采样条件以及如何触发它们、如何更好地控制覆盖组，以及如何减少不关心的覆盖率数据，都是接下来围绕覆盖率应用进行论述的各项疑难点。

2.6.1 covergroup 的采样事件如何指定？

可以在定义 covergroup 时，通过指定采样事件来使能覆盖组的自动采样功能，例如 *covergroup g1 @(posedge clk)*；也可以通过调用 covergroup 的预定义函数 sample() 来完成数据采样。

关键词：
sampling event 采样事件，sample()

避坑指南：
如果覆盖组已经指定了自动采样事件即@event，那么就不需要再额外调用 sample() 函数。

参考代码： sv_covergroup_sample.sv

```
module tb;
  bit[3:0] var1, var2;
  event sample_e;

  covergroup covg1(ref bit[3:0] d1, ref bit[3:0] d2) @sample_e;
    coverpoint d1;
    coverpoint d2;
  endgroup

  covergroup covg2 with function sample(bit[3:0] d1, bit[3:0] d2);
    coverpoint d1;
    coverpoint d2;
  endgroup

  initial begin
    covg1 cg1 = new(var1, var2);
    covg2 cg2 = new();
    fork
      forever #10ns -> sample_e;
      forever begin
        @sample_e;
        var1++;
        var2++;
      end
      forever begin
```

```
          @sample_e;
          cg2.sample(var1, var2);
          if(cg1.get_coverage() == 100 && cg2.get_coverage() == 100) begin
            $display("@%0t :coverge reached 100%%", $time);
            $finish();
          end
        end
      join
    end
endmodule
```

仿真结果：

```
@160000 :coverge reached 100%
```

阅读手记：

2.6.2 covergroup 如何对变量进行采样？

可以在定义 covergroup 时指定要跟踪采样的变量，例如 covergroup cg (ref int x , ref int y)；也可以通过在定义 covergroup 时覆盖预定义 sample()函数为其指定采样的参数，例如 with function sample(bit a, int x)。

关键词：

sampling variables   采样变量，sample()，ref

避坑指南：

如果是作为 covergroup 采样的变量参数，那么应声明为 ref（体现跟踪外部变量的意图）；如果是作为 sample()函数的变量参数，那么则无须声明 ref，因为该参数在每次调用 sample()时会复制外部变量的值。

阅读手记：

2.6.3 是否可对 covergroup 中的不同 coverpoint 指定采样条件？

无论是指定自动的采样事件还是外部调用 sample()函数，默认情况下，covergroup 中每个 coverpoint 均会在该事件进行数值采样。我们可以在定义每一个 coverpoint 时为其单独声明条件，如 *coverpoint s0 iff(!reset)*，该条件表示在采样事件发生的同时，reset 也应该释放，这样才会采样变量 s0。

关键词：
coverpoint，sample()，sampling condition　采样条件

避坑指南：
可灵活使用 iff 采样条件，结合采样事件 *@event iff...* 和 coverpoint 额外条件 *iff...* 来组织更立体的数据采样结构，从而达到灵活高效的采样目的。

参考代码： sv_covergroup_iff.sv

```
module tb;
  bit[3:0] var1, var2;
  bit flag0, flag1, flag2;
  event sample_e;

  covergroup covg1(ref bit[3:0] d1, ref bit[3:0] d2) @(sample_e iff flag0);
    coverpoint d1 iff(flag1);
    coverpoint d2 iff(flag2);
  endgroup

  covergroup covg2 with function sample(bit[3:0] d1, bit[3:0] d2);
    coverpoint d1 iff(flag1);
    coverpoint d2 iff(flag2);
  endgroup

  initial begin
    covg1 cg1 = new(var1, var2);
    covg2 cg2 = new();
    fork
      forever #10ns -> sample_e;
      begin
        #100ns flag0 = 1;
        #100ns flag1 = 1;
        #100ns flag2 = 1;
      end
      forever begin
        @sample_e;
        var1++;
        var2++;
      end
      forever begin
        @(sample_e iff flag0);
        cg2.sample(var1, var2);
        if(cg1.get_coverage() == 100 && cg2.get_coverage() == 100) begin
          $display("@%0t :coverge reached 100%%", $time);
```

```
            $finish();
          end
        end
      join
    end
endmodule
```

仿真结果：

```
@450000 :coverge reached 100%
```

阅读手记：

2.6.4 covergroup 在哪里定义和例化更合适？

一般将 covergroup 定义在 package、class 和 interface 中较为合适，然后将它们在 class 或 interface 中例化。covergroup 的定义和例化的逻辑类似于 assertion，这是由于它们本身有较为独立的数据结构和目标。即便验证环境在向上集成过程中被设置为 PASSIVE，covergroup 的数据采集功能仍然有效。

比如，在 class 中（例如，独立的 coverage model 组件），有利于对接从各个 monitor 连接过来的 analysis port，通过 write() 函数中对应的参数直接对数据做采样。或者，利用 event 触发来做数据采样。

也可以将原本处于 covergroup model 中的 covergroup 实例都搬迁到 interface 中。coverage model 只要有接口句柄（virtual interface），就可以访问 covergroup 实例做数据采样，为此要做的重构工作量也不大。

将 covergroup 实例都搬迁到接口中，也便于其他组件在必要的时候访问 covergroup 实例的数据，或者在上层集成后，由其他组件获取接口句柄，继续做一部分数据采样工作。

关键词：

covergroup，coverage model，interface，class

避坑指南：

记得要例化 covergroup，然后再调用其 sample() 函数。但是，如果 covergroup 定义在了 class 中，那么它不同于 covergroup 定义在 interface、package、module 中的地方是其作为内嵌（embedded）方式定义的，即在定义时，covergroup 的类型为匿名类型（anonymous），而原本的类型名则变为了变量名，因此该变量名也需要在类的 new() 函数中直接例化。

此外，在类中内嵌定义的 covergoup 例化只允许在类的 *new()* 函数中例化。这一点不同之处，往往会给不少验证工程师带来使用上的困惑。同时要注意，这一点限制，也使得同一个内嵌 covergroup 无法在这个类中例化多个实例。

那么，如果想要在类中对同一个covergroup例化多次，建议先将covergroup定义在package中，然后在类中声明并例化多个covergroup实例。

参考代码：sv_covergroup_location.sv

```
package pkg;
  covergroup cg_pkg;
  endgroup
  class covmodel;
    covergroup cg_class;
    endgroup
    function new();
      cg_class = new();
    endfunction
    function init();
      cg_pkg cg_p1 = new();
    endfunction
  endclass
endpackage

interface intf;
  covergroup cg_intf;
  endgroup
  initial begin
    cg_intf cg_i1 = new();
    pkg::cg_pkg cg_p2 = new();
  end
endinterface

module tb;
  covergroup cg_mod;
  endgroup
  initial begin
    cg_mod cg_m1 = new();
  end
endmodule
```

阅读手记：

2.6.5 如果covergroup中的bins没有被采样，可能有哪些原因？

一般我们会将covergroup定义在独立的coverage model或interface中，然后通过analysis imp 端口所处组件实现的 *write()* 函数，获得 monitor 监测到的 transaction 数据，再进行解析和采样。在这一过程中涉及若干逻辑，可以给 covergroup 缺少采样数据的情形提供调试思路。

首先检查 monitor 的 analysis port 与 coverage model 的 analysis imp 是否有连接（connect），再检查 monitor 是否通过该 analysis port 调用 *write()* 函数。接下来，在 coverage model 的 *write()* 函数内部设置某些运行时的断点，检查运行时是否进入该函数体，是否在该函数体内触发了某些逻辑去调用目标 covergroup 的 *sample()* 函数，或触发某些事件以使 covergroup 自动采样（基于事件）。

关键词：
ref，sample，write，connect

避坑指南：
在采样时也应该检查 covergroup 采样的目标变量是否为期望的，例如，其采样的句柄所指向对象的成员可以通过断点设置来查看。又如，一些 covergroup 会通过添加参数来实现采样的灵活化。

那么，如果是基于事件采样（而非 *sample()* 函数和其参数）的，在添加参数列表时要注意，列表中的参数类型若为普通数值类型（例如 logic、int、struct 等），为了能够时刻捕捉到变量的变化，应该将其声明为 ref 类型。否则，input 类型只会在 covergroup 例化时做从外到内的数值复制，但并不会在接下来的数值采样时获得外部变量的数值，这也是 covergroup 采样容易出错的一个地方。

参考代码： V3 验证课程代码

```
// Coverage Model
class rkv_i2c_cgm extends uvm_component;

  // Analysis import declarion below
  uvm_analysis_imp_apb_master
    #(lvc_apb_transfer, rkv_i2c_cgm) apb_trans_observed_imp;

  ...
  covergroup target_address_and_slave_address_cg with function
    sample(bit[9:0]addr, string field);
    option.name = "target_address_and_slave_address_cg";
    TAR_BITS10: coverpoint addr[9:7] iff(field == "TAR"){
      wildcard bins range1 = {3'b1xx};
      wildcard bins range2 = {3'b0xx};
    }

    TAR_BITS7: coverpoint addr[6:0] iff(field == "TAR"){
      wildcard bins range1 = {7'b1xx_xxxx};
      wildcard bins range2 = {7'b0xx_xxxx};
    }

    SAR_BITS10: coverpoint addr[9:7] iff(field == "SAR"){
```

```
      wildcard bins range1 = {3'b1xx};
      wildcard bins range2 = {3'b0xx};
    }

    SAR_BITS7: coverpoint addr[6:0] iff(field == "SAR"){
      wildcard bins range1 = {7'b1xx_xxxx};
      wildcard bins range2 = {7'b0xx_xxxx};
    }
  endgroup
  ...
  virtual function void write_apb_master(lvc_apb_transfer tr);
    uvm_reg r;
    if(tr.trans_status == lvc_apb_pkg::ERROR) return;
    r = cfg.rgm.default_map.get_reg_by_offset(tr.addr);
    ...
    if(r.get_name() == "IC_TAR") begin
     target_address_and_slave_address_cg.sample(
       rgm.IC_TAR_IC_TAR.get(), "TAR");
    end
    else if(r.get_name() == "IC_SAR") begin
      target_address_and_slave_address_cg.sample(
        rgm.IC_SAR_IC_SAR.get(), "SAR");
    end
    ...
  endfunction: write_apb_master
  ...
Endclass
```

阅读手记：

2.6.6 如何减少不关心的 cross bins 采样数据？

就目前较新版本的仿真器如 VCS、Questa 而言，如果用户只对关心的 coverpoint 和 bins 感兴趣，而且在 cross 覆盖率定义时也只对目标 bins 的交叉覆盖率感兴趣，那么有两种办法可用。

一个办法是目前 SystemVerilog 语法支持的，可以在原有 coverpoint 部分将 *option.weight* 设置为 0，再在 cross 部分做之前每个 coverpoint 中单独 bins 的二次重复定义，以及对感兴趣的 bins 做交叉覆盖。这一办法虽然可行，但较为烦琐（从代码维护性和可读性来看不够理想）。

另一个办法，虽然已经从 IEEE-1800 SystemVerilog-2005 标准中开始移除（在 SystemVerilog 3.1a 标准中还存在），但好消息是，目前来看，这个选项的仿真器（以 VCS、

Questa 为例）仍然被支持，那就是在 cross 部分采用 option.cross_auto_bin_max = 0，就能够简单地将可能自动产生的、不感兴趣的（往往是）automatic bins 全部移除，这样就能够获得干净清爽的 cross bins 数据。

关键词：
covergroup，coverpoint，cross，bins

避坑指南：
在仿真器对 cross coverage 还支持 option.cross_auto_bin_max 选项时，使用这个选项可能是一个不那么"守规矩"但好用的选项。

参考代码： sv_cross_cover_bins_efficient.sv

```
module tb;
  logic [1:0] d1, d2;
  covergroup rkv_cg;
    option.per_instance = 1;
    CP_D1: coverpoint d1 {
      option.weight = 0;
      bins v00 = {'b00};
      bins v01 = {'b01};
    }
    CP_D2: coverpoint d2 {
      option.weight = 0;
      bins v10 = {'b10};
      bins v11 = {'b11};
    }

    CRS_D1xD2_weight0: cross CP_D1, CP_D2 {
      bins d1_v00 = binsof(CP_D1.v00);
      bins d1_v01 = binsof(CP_D1.v01);
      bins d2_v10 = binsof(CP_D2.v10);
      bins d2_v11 = binsof(CP_D2.v11);
      bins v00xv10 = binsof(CP_D1.v00) && binsof(CP_D2.v10) ;
      bins v01xv11 = binsof(CP_D1.v01) && binsof(CP_D2.v11) ;
    }

    CRS_D1xD2_automax0: cross CP_D1, CP_D2 {
      option.cross_auto_bin_max = 0;
      bins v00xv10 = binsof(CP_D1.v00) && binsof(CP_D2.v10) ;
      bins v01xv11 = binsof(CP_D1.v01) && binsof(CP_D2.v11) ;
    }

    CRS_d1xd2_automax0: cross d1, d2 {
      option.cross_auto_bin_max = 0;
      bins v00xv10 = (binsof(d1) intersect {'b00}) &&
                     (binsof(d2) intersect {'b10});
      bins v01xv11 = (binsof(d1) intersect {'b01}) &&
```

```
                    (binsof(d2) intersect {'b11});
  }
 endgroup
 initial begin
   rkv_cg cg = new();
   d1 = 'b01;
   d2 = 'b11;
   cg.sample();
   $finish();
 end
endmodule
```

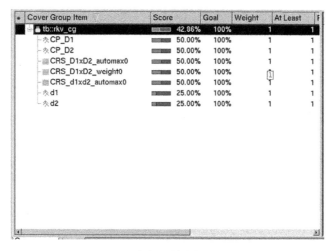

VCS 仿真器覆盖率信息

Questasim 仿真器覆盖率信息

阅读手记：

2.7 线程应用

可以通过 fork 的方式开辟多个并行的子线程，但如何等待子线程执行完毕或终止不再需要的子线程，是需要掌握的地方。子线程之间也有同步和通信的需求，接下来将围绕线程应用讨论一些疑难点。

2.7.1 semaphore 使用时需要初始化吗？

semaphore（旗语）使用时应该例化，即 *semaphore s = new(1)*，表示例化该旗语并给其初始化 1 把钥匙。那么，除了使用 *get(N)* 表示获得 N 把钥匙，我们还可以对已经具备 1 把钥匙的旗语继续归还钥匙吗？是的，理论上，无论谁归还，不管它之前是否"借过"钥匙，都可以采用 *put(N)* 去归还钥匙，而且最终的钥匙总数可以超出它初始化时的数量（尽管这听起来有点不那么"合理"）。

关键词：
semaphore　旗语，get，put

避坑指南：
尽量保持 semaphore 的简单用法，即 1 把钥匙，可供多个组件借还。如果某个组件借到了该钥匙，那么它一定要归还钥匙。当然，在它没有借到钥匙前，它可不能"提前"归还哦（虽然语法允许，但在使用中应当注意）。

参考代码： sv_semaphore_init.sv

```
module tb;
  class rkv_key;
    local semaphore _key;
    local int _key_count;
    local int _key_max;
    function new(int knum = 1);
      _key = new(knum);
      _key_count = knum;
      _key_max = knum;
    endfunction
    function int get_count();
      return _key_count;
    endfunction
    task get_key(int knum = 1);
      _key.get(knum);
      _key_count -= knum;
    endtask
    function bit put_key(int knum = 1);
      if(knum + _key_count <= _key_max) begin
        _key.put(knum);
```

```
        _key_count += knum;
        return 1;
      end
    endfunction
  endclass

  rkv_key key;
  initial begin
    key = new(3);
    $display("key is initialized with count %0d", key.get_count());
    key.get_key(2);
    $display("key current count is %0d after get key num = 2",
             key.get_count());
    for(int i = 3; i > 0; i--) begin
      if(key.put_key(i)) begin
        $display("key current count is %0d after put key num = %0d",
                 key.get_count(), i);
        break;
      end
      else
        $display("key current count %0d and can NOT put key num = %0d",
                 key.get_count(), i);
    end
    $finish();
  end
endmodule
```

仿真结果：

```
key is initialized with count 3
key current count is 1 after get key num = 2
key current count 1 and can NOT put key num = 3
key current count is 3 after put key num = 2
```

阅读手记：

2.7.2 mailbox 使用时需要例化吗？

mailbox（信箱）使用时也应该采用 *new(N)* 函数例化，并且指定其可容纳的成员上限。如果在例化时给定了 *new()* 函数但未传递 *N* 参数，那么该信箱的空间没有上限。在声明信箱变量时，还应尽可能指定该信箱容纳的成员类型，即 *mailbox #(T) m = new(N)*，这样可对其存放的成员类型加以限定，避免因为数据类型放置错误而引起后来的数据读取错误。

关键词：

mailbox，new，type

避坑指南：

信箱的复制（浅拷贝）无法像在例化类的对象时使用 *m2 = new m1* 那样开辟为一个独立的存放空间。需要解释的是，SystemVerilog 标准并没有对这一操作给出规定，以至于不同仿真器对这一操作的行为也不一致（为了代码的可靠性，请谨慎使用 new 的方式做信箱复制）。如果想做信箱内容的复制或合并，那么合适的办法即是采用 *while loop* 做数据成员循环操作，或者使用队列（可能更为方便）。

参考代码： sv_mailbox_copy.sv

```
module tb;
  mailbox #(string) m1, m2;
  string s;
  initial begin
    m1 = new();
    m1.put("cake");
    m1.put("crackers");
    m1.put("cookies");
    m1.put("pie");
    m2 = new m1;
    while(m1.try_get(s)) begin
      $display("m1 content is %p", s);
    end
    while(m2.try_get(s)) begin
      $display("m2 content is %p", s);
    end
  end
endmodule
```

仿真结果：

```
// VCS
m1 content is "cake"
m1 content is "crackers"
m1 content is "cookies"
m1 content is "pie"

// Questa
m1 content is "cake"
m1 content is "crackers"
m1 content is "cookies"
m1 content is "pie"
m2 content is "cake"
m2 content is "crackers"
m2 content is "cookies"
```

```
m2 content is "pie"
```

阅读手记：

2.7.3 fork-join_none 开辟的线程在外部任务退出后也会结束吗？

后台线程不会结束。任何由 fork 开辟的线程（join、join_any、join_none），无论其外部任务（task）何时结束，也无论 fork 何时跳转执行后面的程序，都不会将其开辟的线程自动结束。

对于这些后台的线程，需要考虑它们是否会在接下来影响你的环境（比如，修改信号、变量），或由于反复开辟线程而形成"僵尸"线程。比较安全的办法就是在 fork 满足要求之后，通过变量、旗语或更直接的 disable fork，去终止这些线程。

关键词：

fork，thread，disable fork

避坑指南：

对于有握手、仲裁需求的线程（比如，sequence 发送激励），不应该粗鲁地使用 disable fork 强行结束线程，而应该在线程内部，在其完成握手、仲裁之后的安全阶段结束线程，否则容易引起程序死锁（dead-lock）。

参考代码： sv_fork_join_none_thread_disable.sv

```
module tb;
  task automatic report_time(int n, int id);
    #(n*1ns) $display("@%0t timer[%0d] asserted", $time, id);
  endtask

  task fork_timer;
    fork
      report_time(10, 1);
      report_time(20, 2);
      report_time(30, 3);
    join_none
  endtask

  initial begin
    fork_timer;
    #25ns disable fork;
    $display("@%0t fork threads are disabled", $time);
    $finish();
```

```
        end
    endmodule
```

仿真结果:

```
@10000 timer[1] asserted
@20000 timer[2] asserted
@25000 fork threads are disabled
```

阅读手记:

2.7.4 父线程和子线程之间的执行关系是什么？

父线程用 fork-join_none 开辟子线程后，父线程如果提前结束，而开辟的子线程此时还未结束，那么它还将继续执行。如果在父线程还未结束时，通过 disable 语句来终止父线程，那么子线程也将一同被终止。

关键词:

parent thread　父线程，child thread　子线程，fork-join_none，disable

避坑指南:

父线程即便退出，后台的子线程也将继续执行，所以需要在必要时及时终止子线程，避免后台出现"僵尸"线程。

参考代码: sv_fork_parent_child_thread_disable.sv

```
module tb;
  int s1 = 0;
  int s2 = 0;

  task fork1_t;
    fork
      begin
        #10ns;
        $display("@%0t: time is %0t", $time, $time);
      end
      begin
        wait(s1==1);
        $display("@%0t: disable fork", $time);
        disable fork; // disable current child thread
      end
    join_none
```

```
  endtask

  task fork2_t;
    fork:fork2
      begin: thread2
        #20ns;
        $display("@%0t: time is %t", $time, $time);
      end
      begin
        wait(s2==1);
        $display("@%0t: disable fork2", $time);
        //disable fork; // disable current scope thread
        disable fork2; // disable parent thread and its children threads
        //disable thread2;
      end
    join_none
  endtask

  initial begin
    fork1_t();
    wait fork;
    $display("@%0t: fork1 exited", $time);
  end

  initial begin
    fork2_t();
    wait fork;
    $display("@%0t: fork2 exited", $time);
  end

  initial begin
    #15ns s2 = 1;
  end
endmodule
```

仿真结果:

```
@10000: time is 10000
@15000: disable fork2
@15000: fork2 exited
```

阅读手记:

2.7.5 disable fork 和 disable statement 有什么差别？

disable statement 不仅可以停止 fork 触发的并行线程，还可以停止具名（named）的进程块（process）和方法（task）。disable 可以在该进程或方法的内部来停止该进程或方法，也可以在其他进程或方法停止外部的进程或方法。

这种停止的方法比 disable fork 更为自由。disable fork 停止的是其所在进程或方法中 fork 触发的所有子线程，但它无法停止在它范围以外的子线程。

disable statement 在停止目标时查找的是静态线索，即所有具备该名称的进程或方法；disable fork 在停止目标时的范围是其所在线程及其以下的所有子线程。

关键词：

disable statement，disable fork

避坑指南：

disable statement 可以停止任何具名的进程或方法，但如果在一个系统中具有多个同名进程或方法，那么 disable 将停止所有这些进程或方法。这一点在验证环境出现多个相同组件实例时，很容易出现误杀的情况。可以考虑由 event 触发，在线程内部等待 event，再通过 disable fork 来停止线程，以此避免误杀线程的情况。

参考代码： sv_fork_disable_by_event.sv

```
module tb;
  import uvm_pkg::*;
  `include "uvm_macros.svh"

  class event_data extends uvm_object;
    int unsigned id;
    `uvm_object_utils(event_data)
    function new(string name = "event_data");
    endfunction
  endclass

  task automatic run_time(int t, int id);
    uvm_event stop_e = uvm_event_pool::get_global("stop_e");
    uvm_object tmp;
    event_data ed;
    fork
      #(t*1ns);
      begin
        forever begin
          stop_e.wait_trigger_data(tmp);
          void'($cast(ed, tmp));
          if(id == ed.id) begin
            $display("@%0t:: run_time thread[%0d] is disabled", $time, id);
            disable fork; // disable itself while id matches
          end
        end
```

```
            end
         end
    join_any
    $display("@%0t:: run_time thread[%0d] finished", $time, id);
  endtask

  initial begin
    event_data ed = new("ed");
    uvm_event stop_e = uvm_event_pool::get_global("stop_e");
    for(int i=0; i<5; i++) begin
      automatic int t = i;
      fork
        run_time($urandom_range(10, 20), t);
      join_none
    end
    ed.id = 3;
    #5ns stop_e.trigger(ed); // disable the specified id
    wait fork;
  end
endmodule
```

仿真结果：

```
@5000:: run_time thread[3] is disabled
@11000:: run_time thread[1] finished
@11000:: run_time thread[3] finished
@12000:: run_time thread[4] finished
@17000:: run_time thread[0] finished
@19000:: run_time thread[2] finished
```

阅读手记：

2.7.6 嵌套的 fork 有没有可能被 disable fork 误伤呢？

不会有这种情况。如果能够明确 disable fork 的边界，那么不会误伤到其他相同实例的 fork 线程。但是，如果是按照线程名字使用 disable statement，则并行的多个实例之间就可能会有误伤的情况，那么 disable fork 就有存在的必要性了。

关键词：

disable fork，disable statement

避坑指南：
使用 disable fork 时清楚识别所在的线程边界是正确使用它的关键。

参考代码：sv_fork_disable_embedded.sv

```
module tb;
  task automatic report_time(int n, int id);
    repeat(n) begin
      #1ns $display("timer[%0d] time is %0t", id, $time);
    end
  endtask

  initial begin
    fork
      repeat(3) begin
        fork
          report_time(2, 1);
          report_time(3, 2);
        join_any
        disable fork;
        $display("DISABLE FORK");
      end
    join_none
    wait fork;
    $finish();
  end
endmodule
```

仿真结果：

```
timer[1] time is 1000
timer[2] time is 1000
timer[1] time is 2000
DISABLE FORK
timer[1] time is 3000
timer[2] time is 3000
timer[1] time is 4000
DISABLE FORK
timer[1] time is 5000
timer[2] time is 5000
timer[1] time is 6000
DISABLE FORK
```

阅读手记：

2.7.7 使用 for 配合 fork-join_none 触发多个线程时需要注意什么？

for 循环配合 fork-join_none 可以循环触发多个并行线程，但如果在这些线程之间有数据共用的情况，那么实现的场景很可能不是我们期望的。为了避免在并行线程之间出现数据共用的情况，就要考虑将调用的方法（function、task）及传入的参数变量，都声明为 automatic 属性即动态属性。

在下面的例子中，*run_time_static* 的静态属性使得其变量 d 在多个并行线程之间共享（*run_time_static_proc* 过程块），以及变量 j 在多个并行线程之间共享（*run_time_static_with_auto_var_proc* 过程块），这些都属于容易出错的场景。只要避免并行线程共用静态方法或变量，那么给出的 *run_time_auto_with_auto_var_proc* 过程块就是一个合适的使用实例。

关键词：

for，fork-join_none，static，automatic

避坑指南：

给方法添加 automatic 描述符，仅在 module、interface、package 中需要这么做，对于 class 的方法是不需要额外添加的。但对于避免共用变量的情况，for 循环无论是在 module、interface、package 还是在 class 中，都需要考虑在其内部使用额外的 automatic 变量来暂存外部变量，再将该 automatic 变量作为参数传入 fork-join_none 触发的线程。

参考代码： sv_fork_parallel_thread_run.sv

```systemverilog
module tb;
  task run_time_static (int n, string s = "run_time");
    int d;
    d = n;
    #1ns;
    #(d * 1ns);
    $display("@%0t [%s] finished", $time, s);
  endtask

  task automatic run_time_automatic (int n, string s = "run_time");
    int d;
    d = n;
    #1ns;
    #(d * 1ns);
    $display("@%0t [%s] finished", $time, s);
  endtask

  initial begin
    for(int i = 1; i <= 3; i++) begin : run_time_static_proc
      fork
        run_time_static(i, "run_time1");
      join_none
    end
    wait fork;
```

```
    for(int j = 1; j <= 3; j++) begin : run_time_static_with_auto_var_proc
      automatic int n = j;
      fork
        run_time_static(n, "run_time2");
      join_none
    end
    wait fork;

    for(int k = 1; k <= 3; k++) begin : run_time_auto_with_auto_var_proc
      automatic int n = k;
      fork
        run_time_automatic(n, "run_time3");
      join_none
    end
    wait fork;
  end
endmodule
```

仿真结果：

```
    @5000  [run_time1] finished
    @5000  [run_time1] finished
    @5000  [run_time1] finished
    @9000  [run_time2] finished
    @9000  [run_time2] finished
    @9000  [run_time2] finished
    @11000 [run_time3] finished
    @12000 [run_time3] finished
    @13000 [run_time3] finished
```

阅读手记：

2.8 断言应用

这一节我们将就断言在何处声明和使用以实现更好的复用性，以及通过何种手段对其进行开关控制等疑难点展开讨论。

2.8.1 SV 语言如何控制断言的开关？

可以通过 $asserton(levels [, list])，$assertoff(levels [, list]) 以及相关的其他函数（$assertkill()，$assertcontrol()等）来开关或终止断言。无论这些断言是立即断言还是并行断言，也无论它们属于 assert、assume 或 cover 属性，都可以通过更准确的 $assertcontrol() 对它们的工作状态进行控制。

关键词：
$asserton()，$assertoff()，$assertcontrol()

避坑指南：
除了将断言在复位信号释放之前关闭，还可以在断言属性中将 reset 作为判断条件来关闭断言。如果要对某些范围、模块内的断言做更多实时控制，那么建议将这些控制实现在某些方法中，在仿真过程中进行调用。

参考代码： sv_assert_message_control.sv

```
interface intf;
  bit flag = 0;
  initial begin
    #1ns;
    ast0: assert(flag)
    else $error("flag not true");
  end
endinterface
module tb;
  intf ifi0();
  intf ifi1();
  bit flag = 0;
  initial begin
    #1ns;
    ast0: assert(flag)
    else $error("flag not true");
  end
  initial begin
    $assertoff(0);
    $asserton(0, ifi0);
    $asserton(0, ifi1.ast0);
  end
endmodule
```

仿真结果：

```
Stopping new assertion attempts at time 0ps: level = 0 arg = * (from inst tb
(sv_assert_message_control.sv:19))
Starting assertion attempts at time 0ps: level = 0 arg = tb.ifi0 (from inst tb
(sv_assert_message_control.sv:20))
Starting assertion attempts at time 0ps: level = 0 arg = ifi1.ast0 (from inst tb
(sv_assert_message_control.sv:21))
"sv_assert_message_control.sv", 5: tb.ifi0.ast0: started at 1000ps failed at
1000ps Offending
'flag'
Error: "sv_assert_message_control.sv", 5: tb.ifi0.ast0: at time 1000 ps
flag not true
"sv_assert_message_control.sv", 5: tb.ifi1.ast0: started at 1000ps failed at
1000ps
Offending 'flag'
Error: "sv_assert_message_control.sv", 5: tb.ifi1.ast0: at time 1000 ps
flag not true
```

阅读手记：

2.8.2 仿真器如何控制断言的开关？

这里以 VCS 为例，可以在编译的过程中添加诸如-assert enable_diag -assert enable_hier -assert hier=assert_message_vcs_control.hier 这样的编译选项，控制目标层次中的断言开关。

除了在编译过程中来控制断言，还可以在仿真时控制，比如，在仿真时添加选项-assert hier=assert_message_vcs_run.hier。这么做的好处在于不需要重新编译即可以重新控制断言的开关状态。相比于 SystemVerilog 语言中的开关，这一方式更为灵活。但它仍然是静态方式，无法在仿真过程中适时地开关某些断言。

关键词：

assert on，assert off

避坑指南：

如果想更灵活地在仿真过程中动态地开关断言，可以使用诸如 "*assertion on -force -r -scope tb*" 这样的 tcl 命令行将目标域 tb 下所有的断言都打开。assertion 命令可以添加-*scope ScopeName*，-*module ModuleName* 或-*assert assertion*，以灵活地控制目标断言。

参考代码： sv_assert_message_vcs_control.hier, sv_assert_message_vcs_control.sv, sv_assert_message_vcs_run.hier

```
// sv_assert_message_vcs_control.hier
-assert tb.*
+tree tb.ifi0
+assert tb.ifi1.ast0

// sv_assert_message_vcs_run.hier
-tree tb.*

// sv_assert_message_vcs_control.sv
interface intf;
  bit flag = 0;
  initial begin
    #1ns;
    ast0: assert(flag)
    else $error("flag not true");
  end
endinterface
module tb;
  intf ifi0();
  intf ifi1();
  bit flag = 0;
  initial begin
    #1ns;
    ast0: assert(flag)
    else $error("flag not true");
  end
endmodule
```

仿真结果：

```
"sv_assert_message_vcs_control.sv", 5: tb.ifi0.ast0: started at 0s failed at 0s
  Offending 'flag'
Error: "sv_assert_message_vcs_control.sv", 5: tb.ifi0.ast0: at time 0
flag not true
"sv_assert_message_vcs_control.sv", 5: tb.ifi1.ast0: started at 0s failed at 0s
  Offending 'flag'
Error: "sv_assert_message_vcs_control.sv", 5: tb.ifi1.ast0: at time 0
flag not true
```

阅读手记：

2.8.3 断言在哪里定义和例化更为合适？

首先将断言区分为立即断言和并行断言。SystemVerilog 对立即断言触发的限制不多，可以在常见的 module、interface、class 中触发。

并行断言的限制比立即断言要多一些。SystemVerilog 不支持在类中（动态实例）例化并行断言（静态属性）。所以对于更为常见的并行断言，合适的处理方法是将公用的断言库（sequence、property）定义在 package 中，而在 interface、module 中去定义与设计功能相关的 sequence、property，并且去例化它们。

断言的独立性较高，更适合在接口中定义和例化它们。这样，独立的接口往往可以用来嵌入、绑定到设计中，做更为灵活的时序、功能检查。

关键词：

immediate，concurrent，property，package，interface，module

避坑指南：

配合验证环境组件中对于某些功能、时序的检查任务，还需要考虑哪些检查适合采用任务（task）来完成，哪些检查适合采用并行断言来完成。如果不考虑形式验证对断言复用的情况，那么可以将复杂的、长的时序用 task 来完成，清晰的、短的时序用 property 来完成。当然，复杂的时序也可以拆分为多个 property。另外也要注意，如果希望通过 UVM 报告的方式来统一管理断言的消息，请在 *assert property()* 后添加 *else `UVM_ERROR()* 的错误消息报告。

参考代码： sv_assertion_location.sv

```
package pkg;
  property prop2;
    1;
  endproperty
  //ast_prop2: assert property(prop2); // ERROR location

  class chk;
    // ast_prop2: assert property(pkg::prop2); // ERROR location
    task check();
      //ast_prop2: assert property(pkg::prop2); // ERROR location
      ast_imm2: assert(1);
    endtask
  endclass
endpackage

interface intf;
  import uvm_pkg::*;
  `include "uvm_macros.svh"
  property prop1;
    1;
  endproperty
  ast_prop1: assert property(prop1) else `uvm_error("ASSERT", "ERROR")
  ast_prop2: assert property(pkg::prop2)
```

```
                else `uvm_error("ASSERT", "ERROR")
  initial begin
    ast_imm1: assert(1);
  end
endinterface
```

阅读手记：

2.8.4 如何更好地让接口中的断言实现复用性？

应该考虑到接口中的断言的独立性、可控性，以及在仿真验证和形式验证中的移植性。

独立性在于接口中的断言涉及的信号、变量只应该存在于该接口中，避免与外部变量形成耦合。同时，这些断言的"尺寸"应该朝着小型化靠拢，这样不仅易于阅读、维护，也便于断言出错后的调试。

可控性在于接口中的所有断言对复位信号保持敏感，也提供可以随时控制该接口中的断言（如开关断言）的方法。

移植性的缺失，可能是多数验证工程师不会在形式验证中应用同一接口中断言的原因。如果要将这些断言也应用在形式验证中，就得考虑添加参数，利用 generate 块，在不同参数配置情况下，对同一个 property 例化时采取 assert 还是 assume 的方式。

关键词：

assertion，interface，parameter，generate，assert，assume

避坑指南：

在形式验证中，接口中的同一个断言，在验证 slave 或 master 时需通过参数分配选择 assert 或 assume 的形式。不过，在动态仿真中，assert 和 assume 都将按照 assert 的行为做断言检查。

cover property 的形式在动态仿真和形式验证中都将参与到覆盖率收集中。如果采取的是同一家公司的 EDA 工具链，那么动态仿真和形式验证收集的覆盖率数据还可以按照相同的覆盖率数据格式做最后的合并，体现出更完整的验证覆盖数据。

参考代码：sv_assertion_sim_formal_reuse.sv

```
package rkv_pkg;
  property prop2;
    1;
  endproperty
endpackage

interface intf;
```

```
      parameter bit verify_master = 1;
      property prop1;
        1;
      endproperty
      generate
        if(verify_master) begin
          ast_prop1: assert property(prop1);
          ast_prop2: assert property(rkv_pkg::prop2);
        end
        else begin
          asm_prop1: assume property(prop1);
          asm_prop2: assume property(rkv_pkg::prop2);
        end
      endgenerate
    endinterface
```

阅读手记：

第 3 章

UVM 疑难点集合

3.1 UVM 机制

从使用角度来看 UVM 的机制，如果遵循它的常规建议，那么出现问题的可能性会比较少。在实际项目中，往往需要结合 UVM 机制做一些相对复杂的、超常规的操作。比如，对于对象的创建和 phase 执行顺序问题、消息等级设定和过滤问题、延时退出问题、变量传递和配置问题等，构成了一些使用中的疑难点。接下来，我们将就这些问题展开讨论。

3.1.1 是否所有的 UVM 对象都应该用工厂创建呢？

建议初学者在创建 UVM 对象时，无论是 uvm_component 类型（和子类）还是 uvm_object 类型（和子类），都应该用工厂（factory）去创建。用工厂创建 UVM 对象的方法有很多，常见的是调用 T::type_id::create(name, parent) 函数。利用工厂创建 UVM 对象带来的直接好处就是便于上层集成环境层（environment layer）或测试层（test layer）对目标类型进行类型覆盖（type override）或实例覆盖（instance override）。

关键词：
object creation 对象创建，factory 工厂，override 覆盖

避坑指南：
在覆盖目标对象前，确保该对象由工厂创建，且对覆盖函数的调用应该发生在创建对象之前。

参考代码： uvm_factory_create_override.sv

```
package rkv_pkg;
  import uvm_pkg::*;
  `include "uvm_macros.svh"
  class rkv_comp extends uvm_component;
    `uvm_component_utils(rkv_comp)
    function new(string name = "rkv_comp", uvm_component parent = null);
      super.new(name, parent);
    endfunction
  endclass
```

```systemverilog
    class rkv_comp_x extends rkv_comp;
      `uvm_component_utils(rkv_comp_x)
      function new(string name = "rkv_comp_x", uvm_component parent = null);
        super.new(name, parent);
      endfunction
    endclass
    class rkv_comp_y extends rkv_comp;
      `uvm_component_utils(rkv_comp_y)
      function new(string name = "rkv_comp_y", uvm_component parent = null);
        super.new(name, parent);
      endfunction
    endclass
    class factory_create_override_test extends uvm_test;
      rkv_comp comp;
      `uvm_component_utils(factory_create_override_test)
      function new(string name = "factory_create_override_test",
                   uvm_component parent = null);
        super.new(name, parent);
      endfunction
      function void build_phase(uvm_phase phase);
        super.build_phase(phase);
        // available to override
        set_type_override("rkv_comp", "rkv_comp_x");
        comp = rkv_comp::type_id::create("comp", this);
        // too late after instance created
        set_type_override("rkv_comp", "rkv_comp_y");
        `uvm_info("TYPE",
          $sformatf("comp type name is %s", comp.get_type_name()), UVM_LOW)
      endfunction
    endclass
endpackage

module tb;
  import uvm_pkg::*;
  `include "uvm_macros.svh"
  import rkv_pkg::*;
  initial run_test("factory_create_override_test");
endmodule
```

仿真结果：

```
      UVM_INFO @ 0: reporter [RNTST] Running test factory_create_override_
test...
      UVM_INFO @ 0: reporter [TPREGR] Original object type 'rkv_comp' already
registered to produce 'rkv_comp_x'.  Replacing with override to produce type
'rkv_comp_y'.
      UVM_INFO uvm_factory_create_override.sv(35) @ 0: uvm_test_top [TYPE] comp
type name is rkv_comp_x
```

阅读手记：

3.1.2 工厂创建 uvm_object 是否需要为其指定 parent？

严格而言，工厂在创建 uvm_object 时，如果其所处对象的类型为 uvm_component 类型（或子类），则可以将该组件实例指定为其 parent；如果创建 uvm_object 的上层也为 uvm_object 类型（非 uvm_component 类型），则不需要指定 T::type_id::create(name, parent=null)的第二个参数，即采用默认值 null。如此，该例化的对象在 UVM 结构配置层次（用作 uvm_config_db::{set, get}）中以 uvm_root（最顶层）为其 parent。

关键词：
uvm_object，uvm_component，create，parent，uvm_config_db

避坑指南：
在创建 uvm_object 对象时，一般不需要指定第二个参数，除非你确定需要将该 uvm_object 对象也置于 UVM 结构配置层次下，并利用该层次通过 uvm_config_db 设置和获取某些配置变量。这种采取默认值的固定创建方式便于记忆。

参考代码： uvm_object_create_under_parent.sv

```
package rkv_pkg;
  import uvm_pkg::*;
  `include "uvm_macros.svh"
  class rkv_object extends uvm_object;
    `uvm_object_utils(rkv_object)
    function new(string name = "rkv_object");
      super.new(name);
    endfunction
  endclass
  class rkv_config extends uvm_object;
    rkv_object obj;
    `uvm_object_utils(rkv_config)
    function new(string name = "rkv_config");
      super.new(name);
    endfunction
  endclass
  class rkv_comp extends uvm_component;
    rkv_config cfg;
    `uvm_component_utils(rkv_comp)
    function new(string name = "rkv_comp", uvm_component parent = null);
      super.new(name, parent);
```

```
    endfunction
    function void build_phase(uvm_phase phase);
      // cfg is attached under current component and
      // available for config_db set/get pair
      cfg = rkv_config::type_id::create("cfg", this);
      if(!uvm_config_db#(rkv_object)::get(this, "cfg", "obj", cfg.obj))
        `uvm_error("GETCFG","cannot get rkv_object handle from config DB")
      else
        `uvm_info("GETCFG",
          "got rkv_object handle from config DB", UVM_LOW)
    endfunction
  endclass
  class object_create_under_parent_test extends uvm_test;
    rkv_object obj;
    rkv_comp comp;
    `uvm_component_utils(object_create_under_parent_test)
    function new(string name = "object_create_under_parent_test",
              uvm_component parent = null);
      super.new(name, parent);
    endfunction
    function void build_phase(uvm_phase phase);
      obj = rkv_object::type_id::create("obj", this);
      uvm_config_db#(rkv_object)::set(this, "comp.cfg","obj", obj);
      comp = rkv_comp::type_id::create("comp", this);
    endfunction
  endclass
endpackage

module tb;
  import uvm_pkg::*;
  `include "uvm_macros.svh"
  import rkv_pkg::*;
  initial run_test("object_create_under_parent_test");
endmodule
```

仿真结果：

```
UVM_INFO @ 0: reporter [RNTST] Running test
object_create_under_parent_test...
UVM_INFO uvm_object_create_under_parent.sv(30) @ 0: uvm_test_top.comp
[GETCFG] got rkv_object handle from config DB
```

阅读手记：

3.1.3 为什么建议配置放在对象创建之前？

这是为了将所需配置的变量事先放入 uvm_config_db 中进行存储，而在后续子一级组件例化并且进入 build_phase 从 uvm_config_db 获得配置变量时，可以确保所配置的变量都已先于子一级组件创建前进行过配置，避免出现配置变量无法获取的情况。

关键词：
uvm_config_db，create，build_phase

避坑指南：
在 build_phase 中，对当前层次及以下组件的变量配置，都应该先于该子一级组件创建前完成。尽管实际上子一级组件先创建而后再对子一级组件做相关变量配置的做法，往往也可以成功，但为了避免可能的麻烦，我们仍然建议将配置放在对象创建之前。

阅读手记：

3.1.4 UVM 环境中进入 new()和 build_phase()有什么区别？

为了简单起见，除了对成员变量的初始化可以放在 *new()* 函数中。其他诸如对象创建、类型覆盖、变量配置，我们都应该放入到 *build_phase()* 函数中。之所以这样做，主要是为了让 UVM 环境中的对象（uvm_object、uvm_component）在创建之前，工厂可以检查类型是否被覆盖，并且选择正确的类型进行创建。而如果将创建对象的步骤放入 *new()* 函数中，那么将可能失去这个机会。换句话说，如果所有的对象创建都在 *new()* 函数中，那么对象的创建将不再直接参与到 build 阶段，因此不能很好地配合类型覆盖机制，继而无法选择正确的类型去创建对象。

关键词：
new，build_phase，override

避坑指南：
为了配合类型覆盖机制，在创建被注册到工厂的类型时，应该使用 *T::type_id::create()* 的方式去创建对象，即利用工厂去创建对象（封装了"先选择类型，后创建对象"的执行逻辑）。

参考代码： uvm_create_build_or_new.sv

```
package rkv_pkg;
  import uvm_pkg::*;
  `include "uvm_macros.svh"
  class rkv_comp extends uvm_component;
```

```systemverilog
      `uvm_component_utils(rkv_comp)
      function new(string name = "rkv_comp", uvm_component parent = null);
        super.new(name, parent);
      endfunction
      function void end_of_elaboration_phase(uvm_phase phase);
        `uvm_info("TYPE",
          $sformatf("[%s] type name is [%s]", get_name(), get_type_name()),
          UVM_LOW)
      endfunction
    endclass
    class user_comp extends rkv_comp;
      `uvm_component_utils(user_comp)
      function new(string name = "user_comp", uvm_component parent = null);
        super.new(name, parent);
      endfunction
    endclass
    class create_build_or_new_test extends uvm_test;
      rkv_comp comp0, comp1;
      `uvm_component_utils(create_build_or_new_test)
      function new(string name = "create_build_or_new_test",
                uvm_component parent = null);
        super.new(name, parent);
        comp0 = rkv_comp::type_id::create("comp0", this);
      endfunction
      function void build_phase(uvm_phase phase);
        super.build_phase(phase);
        // available to override
        set_type_override("rkv_comp", "user_comp");
        comp1 = rkv_comp::type_id::create("comp1", this);
      endfunction
    endclass
  endpackage

  module tb;
    import uvm_pkg::*;
    `include "uvm_macros.svh"
    import rkv_pkg::*;
    initial run_test("create_build_or_new_test");
  endmodule
```

仿真结果：

```
       UVM_INFO @ 0: reporter [RNTST] Running test create_build_or_new_test...
       UVM_INFO uvm_create_build_or_new.sv(10) @ 0: uvm_test_top.comp0 [TYPE]
[comp0] type name is [rkv_comp]
       UVM_INFO uvm_create_build_or_new.sv(10) @ 0: uvm_test_top.comp1 [TYPE]
[comp1] type name is [user_comp]
```

阅读手记：

3.1.5 在创建组件时采用 new() 有什么影响？

可以肯定的是，在创建组件时采用了 *new()* 函数，不会影响该组件的各个 phase 阶段的执行，如 build_phase、connect_phase。各个 phase 阶段的执行依赖于 parent-child component 层次结构，而该结构的形成归结于 *uvm_component::new()* 函数。但是，如果创建组件时采用了 *T::type_id::create()* 函数，那么它会间接调用 *new()* 函数，在此之前还会检查目标类型是否被覆盖（override）。

关键词：

new，create，phase，override

避坑指南：

为了便于接下来的类型覆盖，在创建被注册过的类型时，尽量使用 *T::type_id::create()* 函数而不使用 *new()* 函数，因为后者无法支持类型覆盖。

参考代码： uvm_object_create_diff_new.sv

```
package rkv_pkg;
  import uvm_pkg::*;
  `include "uvm_macros.svh"
  class rkv_comp extends uvm_component;
    `uvm_component_utils(rkv_comp)
    function new(string name = "rkv_comp", uvm_component parent = null);
      super.new(name, parent);
    endfunction
    function void end_of_elaboration_phase(uvm_phase phase);
      `uvm_info("TYPE",
        $sformatf("%s is created by type %s", get_name(), get_type_name()),
        UVM_LOW)
    endfunction
  endclass
  class user_comp extends rkv_comp;
    `uvm_component_utils(user_comp)
    function new(string name = "user_comp", uvm_component parent = null);
      super.new(name, parent);
    endfunction
  endclass
  class object_create_diff_new_test extends uvm_test;
    rkv_comp c0, c1;
```

```
    `uvm_component_utils(object_create_diff_new_test)
    function new(string name = "object_create_diff_new_test",
                 uvm_component parent = null);
      super.new(name, parent);
    endfunction
    function void build_phase(uvm_phase phase);
      set_type_override("rkv_comp", "user_comp");
      c0 = new("c0", this);
      c1 = rkv_comp::type_id::create("c1", this);
    endfunction
  endclass
endpackage

module tb;
  import uvm_pkg::*;
  `include "uvm_macros.svh"
  import rkv_pkg::*;
  initial run_test("object_create_diff_new_test");
endmodule
```

仿真结果：

```
UVM_INFO @ 0: reporter [RNTST] Running test object_create_diff_new_test...
UVM_INFO uvm_object_create_diff_new.sv(10) @ 0: uvm_test_top.c0 [TYPE]
c0 is created by type rkv_comp
UVM_INFO uvm_object_create_diff_new.sv(10) @ 0: uvm_test_top.c1 [TYPE]
c1 is created by type user_comp
```

阅读手记：

3.1.6 UVM 配置类的参数修改应该发生在什么时间？

对于一些复杂的、需要配置的 VIP（Verification IP），会在顶层 build_phase 阶段例化且配置这些 VIP 的配置类（configuration class），这将决定接下来 VIP 在仿真中的行为，使其能够按照特定的协议要求，或与 DUT 的功能配置保持一致。有些 VIP 也支持在 run_phase 阶段进行再配置（reconfiguration），继而与设计的功能变化保持协调。

关键词：

VIP，configuration class 配置类，reconfiguration 再配置，build_hase

避坑指南：

在应用 VIP 的再配置功能时，应该注意，哪些参数可以在 run_phase 中再配置、哪些参数不支持再配置。

阅读手记：

3.1.7 UVM 的消息严重等级是否可以屏蔽或修改触发动作？

我们可以使用函数将消息的 ID 或 ID-SEVERITY 进行匹配对应，并且指定 uvm_action 参数为 UVM_NO_ACTION，即可将原本要打印（UVM_DISPLAY）的消息 ID 屏蔽。但同时应注意，对消息等级的设定，应该在顶层 test 的 *end_of_elaboration_phase()* 函数中完成。因为在所有组件层次结构确定后，可以通过以下相关函数设定不同层次组件的消息 ID 的触发动作。

关键词：
report，severity

避坑指南：
如果消息 ID 是从 uvm_object 中发出的，则需要从最顶层（uvm_root）对消息 ID 进行管理。例如，*uvm_root::get().set_report_id_action_hier()*，这种方式可以管理 object 对象和 component 组件发出的消息。

参考代码： uvm_component.svh（UVM 源代码）

```
function void set_report_id_action (string id, uvm_action action);
function void set_report_id_action_hier (string id, uvm_action action);
function void set_report_severity_id_action (uvm_severity severity,
                                             string id,
                                             uvm_action action);
function void set_report_severity_id_action_hier(uvm_severity severity,
                                                 string id,
                                                 uvm_action action);
```

阅读手记：

3.1.8 UVM 的消息严重等级是否可以修改？

UVM 提供可以修改消息严重级别（SEVERITY）的命令，只不过有点可惜的是，它不提供层次化的递归设置，即缺少后缀名为 *_hier(xxx)* 这样的函数。假如要将以下示例中 c1 和 c1.c2

的消息 ID 为 "RUN" 的 WARNING 信息级别修改为 INFO 级别，那么需要调用：

  c1.set_report_severity_id_override(UVM_WARNING, "RUN", UVM_INFO);
  c1.c2.set_report_severity_id_override(UVM_WARNING, "RUN", UVM_INFO);

关键词：
report，severity，override

避坑指南：
遗憾的一点是，UVM 确实缺少类似以下用作消息严重等级递归设置的函数，不过该函数可以由用户来实现。

参考代码： uvm_message_severity_surpress.sv

```
package rkv_pkg;
  import uvm_pkg::*;
  `include "uvm_macros.svh"
  class rkv_comp2 extends uvm_component;
    `uvm_component_utils(rkv_comp2)
    function new(string name = "rkv_comp2", uvm_component parent = null);
      super.new(name, parent);
    endfunction
    task run_phase(uvm_phase phase);
      super.run_phase(phase);
      `uvm_warning("RUN", "rkv_comp2 run phase entered")
      `uvm_warning("RUN", "rkv_comp2 run phase exited")
    endtask
  endclass

  class rkv_comp1 extends uvm_component;
    rkv_comp2 c2;
    `uvm_component_utils(rkv_comp1)
    function new(string name = "rkv_comp1", uvm_component parent = null);
      super.new(name, parent);
    endfunction
    function void build_phase(uvm_phase phase);
      super.build_phase(phase);
      c2 = rkv_comp2::type_id::create("c2", this);
    endfunction
    task run_phase(uvm_phase phase);
      super.run_phase(phase);
      `uvm_warning("RUN", "rkv_comp1 run phase entered")
      `uvm_warning("RUN", "rkv_comp1 run phase exited")
    endtask
  endclass

  class message_severity_surpress_test extends uvm_test;
    rkv_comp1 c1;
    rkv_comp2 c2;
```

```
`uvm_component_utils(message_severity_surpress_test)
function new(string name = "message_severity_surpress_test",
             uvm_component parent = null);
  super.new(name, parent);
endfunction
function void build_phase(uvm_phase phase);
  super.build_phase(phase);
  c1 = rkv_comp1::type_id::create("c1", this);
  c2 = rkv_comp2::type_id::create("c2", this);
endfunction

function void end_of_elaboration_phase(uvm_phase phase);
  // available to override severity for specific component
//c1.set_report_severity_id_override(UVM_WARNING, "RUN", UVM_INFO);
//c1.c2.set_report_severity_id_override(UVM_WARNING, "RUN", UVM_INFO);

  // available to override severity hierarchically
  set_report_severity_id_override_hier(UVM_WARNING,
                                       "RUN", UVM_INFO);
endfunction

function void set_report_severity_id_override_hier(
                                  uvm_severity cur_severity,
                                  string id,
                                  uvm_severity new_severity,
                                  int depth = 10
                                  );
  uvm_component children[$];
  m_rh.set_severity_id_override(cur_severity, id, new_severity);
  get_depth_children(this, children, depth);
  foreach(children[i])
    children[i].m_rh.set_severity_id_override(cur_severity,
                                              id, new_severity);
endfunction

function void get_depth_children(input uvm_component h,
                                 ref uvm_component children[$],
                                 input int depth=1
                                 );
  if(depth > 0) begin
    foreach(h.m_children[i]) begin
      children.push_back(h.m_children[i]);
      get_depth_children(h.m_children[i], children, depth-1);
    end
  end
endfunction
endclass
```

```
            endpackage

            module tb;
              import uvm_pkg::*;
              `include "uvm_macros.svh"
              import rkv_pkg::*;
              initial run_test("message_severity_surpress_test");
            endmodule
```

仿真结果：

```
        UVM_INFO @ 0: reporter [RNTST] Running test
        message_severity_surpress_test...
        UVM_INFO uvm_message_severity_surpress.sv(11) @ 0: uvm_test_top.c1.c2
[RUN] rkv_comp2 run phase entered
        UVM_INFO uvm_message_severity_surpress.sv(12) @ 0: uvm_test_top.c1.c2
[RUN] rkv_comp2 run phase exited
        UVM_INFO uvm_message_severity_surpress.sv(28) @ 0: uvm_test_top.c1 [RUN]
rkv_comp1 run phase entered
        UVM_INFO uvm_message_severity_surpress.sv(29) @ 0: uvm_test_top.c1 [RUN]
rkv_comp1 run phase exited
        UVM_INFO uvm_message_severity_surpress.sv(11) @ 0: uvm_test_top.c2 [RUN]
rkv_comp2 run phase entered
        UVM_INFO uvm_message_severity_surpress.sv(12) @ 0: uvm_test_top.c2 [RUN]
rkv_comp2 run phase exited
```

阅读手记：

3.1.9 通过 uvm_config_db 可以完成哪些数据类型的配置？

可以通过 *uvm_config_db::{set, get}* 完成以下常见的多种数据类型的配置，包括整型（int, byte, integer 等）、向量（bit[M:N], logic[M:N]）、枚举类型、结构体类型、句柄、虚接口（virtual interface）、字符串、信箱、旗语等。不过较为遗憾的是，我们还无法将各种数组类型（定长数组、动态数组、队列、关联数组）通过 uvm_config_db 来完成配置。

关键词：

uvm_config_db，set，get

避坑指南：

如果需要配置大量的数据内容，或者配置数组内容，那么可以将这些数据封装在一个配置类中，在顶层将该配置类进行例化，通过 uvm_config_db 完成该配置对象的句柄传递，而底层组件则可以通过获得的配置对象的句柄，进一步访问该配置对象中的多个变量。

阅读手记：

3.1.10 使用 uvm_config_db 时传递的参数类型是否需要保持一致？

传递的参数类型需要严格保持一致。由于 *uvm_config_db #(type T)* 是参数类，在使用 *set()* 和 *get()* 方法时，均需要相同的类型进行传递，才能获得所传递的变量。例如，如果上层传递时通过 *#(T=ENUM)* 指定枚举类型为参数，那么底层应该也以枚举类型的参数形式 *#(T=ENUM)* 来获取，而不应该使用诸如 int 这样其他的类型参数来试图获取；又例如，上层传递时通过 *#(T=SUBCLASS)* 指定子类为参数，那么底层也应该以子类的参数形式 *#(T=SUBCLASS)*，而不应该试图使用诸如父类这样其他的类型参数来获取。

关键词：
uvm_config_db，set，get，type

避坑指南：

SystemVerilog 在不同类型赋值时，会发生隐式类型转换（implicit type conversion），但在使用 uvm_config_db 获得参数传递时，应严格保持参数类型的一致性。简而言之，就是 *uvm_config_db #(type T)* 中的参数类型在传递和获取时应该保持一致。

参考代码： uvm_config_set_get_type.sv

```
interface rkv_intf;
endinterface

package rkv_pkg;
  import uvm_pkg::*;
  `include "uvm_macros.svh"
  typedef enum {WRITE, READ, IDLE} command_t;
  typedef struct {
    string name;
    int num;
  } score_t;
  class rkv_comp extends uvm_component;
    command_t cmd;
    string info;
    score_t score;
    bit[1:0] ctrl;
    virtual rkv_intf vif;
    `uvm_component_utils(rkv_comp)
    function new(string name = "rkv_comp", uvm_component parent = null);
      super.new(name, parent);
```

```systemverilog
    if(!uvm_config_db#(int)::get(this, "", "cmd", cmd))
      `uvm_warning("CFGGET","cannot get cmd from config DB")
    else
      `uvm_info("CFGGET", $sformatf("Got cmd = %s", cmd), UVM_LOW)
    if(!uvm_config_db#(integer)::get(this, "", "cmd", cmd))
      `uvm_warning("CFGGET",
                   "cannot get cmd from config DB via integer type")
    else
      `uvm_info("CFGGET", $sformatf("Got cmd = %s", cmd), UVM_LOW)
    if(!uvm_config_db#(byte)::get(this, "", "cmd", cmd))
      `uvm_warning("CFGGET",
                   "cannot get cmd from config DB via byte type")
    else
      `uvm_info("CFGGET", $sformatf("Got cmd = %s", cmd), UVM_LOW)
    if(!uvm_config_db#(string)::get(this, "", "info", info))
      `uvm_warning("CFGGET","cannot get info from config DB")
    else
      `uvm_info("CFGGET", $sformatf("Got info = %s", info), UVM_LOW)
    if(!uvm_config_db#(score_t)::get(this, "", "score", score))
      `uvm_warning("CFGGET","cannot get score from config DB")
    else
      `uvm_info("CFGGET",
        $sformatf("Got score name = %s, num = %0d", score.name, score.num),
        UVM_LOW)
    if(!uvm_config_db#(bit[1:0])::get(this, "", "ctrl", ctrl))
      `uvm_warning("CFGGET","cannot get score from config DB")
    else
      `uvm_info("CFGGET", $sformatf("Got ctrl = 'b%0b", ctrl), UVM_LOW)
    if(!uvm_config_db#(bit[3:0])::get(this, "", "ctrl", ctrl))
      `uvm_warning("CFGGET",
                   "cannot get ctrl from config DB via bit[3:0] type")
    else
      `uvm_info("CFGGET", $sformatf("Got ctrl = 'b%0b", ctrl), UVM_LOW)
    if(!uvm_config_db#(virtual rkv_intf)::get(this, "", "vif", vif))
      `uvm_warning("CFGGET","cannot get vif from config DB")
    else
      `uvm_info("CFGGET", "Got vif", UVM_LOW)
  endfunction
endclass

class config_set_get_type_test extends uvm_test;
  rkv_comp c1;
  string info = "config";
  score_t score = '{name:"check", num:100};
  bit [1:0] ctrl = 2'b01;
  virtual rkv_intf vif;
  `uvm_component_utils(config_set_get_type_test)
```

```
    function new(string name = "config_set_get_type_test",
                 uvm_component parent = null);
      super.new(name, parent);
    endfunction
    function void build_phase(uvm_phase phase);
      super.build_phase(phase);
      uvm_config_db#(int)::set(this, "c1", "cmd", 1);
      uvm_config_db#(string)::set(this, "c1", "info", info);
      uvm_config_db#(score_t)::set(this, "c1", "score", score);
      uvm_config_db#(bit[1:0])::set(this, "c1", "ctrl", ctrl);
      uvm_config_db#(virtual rkv_intf)::set(this, "c1", "vif", vif);
      c1 = rkv_comp::type_id::create("c1", this);
    endfunction
  endclass
endpackage

module tb;
  import uvm_pkg::*;
  `include "uvm_macros.svh"
  import rkv_pkg::*;
  initial run_test("config_set_get_type_test");
endmodule
```

仿真结果：

```
      UVM_INFO @ 0: reporter [RNTST] Running test config_set_get_type_test...
      UVM_INFO uvm_config_set_get_type.sv(24) @ 0: uvm_test_top.c1 [CFGGET] Got
cmd = READ
      UVM_WARNING uvm_config_set_get_type.sv(26) @ 0: uvm_test_top.c1 [CFGGET]
cannot get cmd from config DB via integer type
      UVM_WARNING uvm_config_set_get_type.sv(30) @ 0: uvm_test_top.c1 [CFGGET]
cannot get cmd from config DB via byte type
      UVM_INFO uvm_config_set_get_type.sv(36) @ 0: uvm_test_top.c1 [CFGGET] Got
info = config
      UVM_INFO uvm_config_set_get_type.sv(40) @ 0: uvm_test_top.c1 [CFGGET] Got
score name = check, num = 100
      UVM_INFO uvm_config_set_get_type.sv(44) @ 0: uvm_test_top.c1 [CFGGET] Got
ctrl = 'b1
      UVM_WARNING uvm_config_set_get_type.sv(46) @ 0: uvm_test_top.c1 [CFGGET]
cannot get ctrl from config DB via bit[3:0] type
      UVM_INFO uvm_config_set_get_type.sv(52) @ 0: uvm_test_top.c1 [CFGGET] Got
vif
```

阅读手记：

3.1.11 如何设置 timeout 时间防止仿真超时？

可以利用 *uvm_root::set_timeout(time timeout, bit overridable=1)* 函数来设置。具体的操作可以在 *build_phase()* 或 *end_of_elaboration_phase()* 执行以下语句：

uvm_root::get().set_timeout(TIME)。

关键词：
timeout，uvm_root

避坑指南：
UVM-1.2 之前的全局函数 *set_global_timeout(time timeout, bit overridable = 1)* 应该避免使用，因为该函数在 UVM-1.2 版本中是待弃用的（编译时可采用 +*define+UVM_NO_DEPRECATED* 来弃用 UVM-1.2 版本中的这些代码）。

阅读手记：

3.1.12 set_drain_time()的作用是什么？

我们有时候可以在 *run_phase()* 中，或在 12 个子 phase 如 *main_phase()* 中，见到这样的用法：*phase.phase_done.set_drain_time(this, TIME)*。这里的 TIME 如果设置为 1μs，表示的是在所有相同的 phase 全部完成 drop objection 操作时，再等待 TIME 时间才进入下一个 phase。

最常见的是，如果只利用 *run_phase()* 或 *main_phase()* 在该 phase 中设定 *set_drain_time()*，则表示待完成 drop objection 操作时再等待 TIME 时间，仿真才会进入下一个 phase（往往意味着仿真将要结束）。这种情况可以用来在一些数据流测试中等待所有的数据全部从 DUT 中搬出，而不需要在 sequence 中额外等待某一段固定的时间。

关键词：
set_drain_time，run_phase，main_phase

避坑指南：
set_drain_time() 只应该在 task phase 中使用，例如 *run_phase()*、*main_phase()*。如果在同一个 phase，不同的组件中都使用了 *set_drain_time()*，那么最后等待进入下一个 phase 的额外时间是所有设定的 drain time 之和。此外，还需要将其与 *uvm_root::set_timeout()* 加以区分。

参考代码： uvm_test_drain_time.sv

```
package rkv_pkg;
  import uvm_pkg::*;
  `include "uvm_macros.svh"
  class rkv_comp extends uvm_component;
```

```
  `uvm_component_utils(rkv_comp)
  function new(string name = "rkv_comp", uvm_component parent = null);
    super.new(name, parent);
  endfunction
  task reset_phase(uvm_phase phase);
    `uvm_info("PHASE", "reset phase entered", UVM_LOW)
    phase.raise_objection(this);
    phase.phase_done.set_drain_time(this, 40us);
    #30us;
    phase.drop_objection(this);
    `uvm_info("PHASE", "reset phase exited", UVM_LOW)
  endtask
  task main_phase(uvm_phase phase);
    `uvm_info("PHASE", "main phase entered", UVM_LOW)
    phase.raise_objection(this);
    phase.phase_done.set_drain_time(this, 40us);
    #30us;
    phase.drop_objection(this);
    `uvm_info("PHASE", "main phase exited", UVM_LOW)
  endtask
endclass

class drain_time_test extends uvm_test;
  rkv_comp c1;
  `uvm_component_utils(drain_time_test)
  function new(string name = "drain_time_test",
               uvm_component parent = null);
    super.new(name, parent);
  endfunction
  function void build_phase(uvm_phase phase);
    c1 = rkv_comp::type_id::create("c1", this);
    //uvm_root::get().set_timeout(10us);
  endfunction
  task reset_phase(uvm_phase phase);
    `uvm_info("PHASE", "reset phase entered", UVM_LOW)
    phase.raise_objection(this);
    phase.phase_done.set_drain_time(this, 20us);
    #10us;
    phase.drop_objection(this);
    `uvm_info("PHASE", "reset phase exited", UVM_LOW)
  endtask
  task main_phase(uvm_phase phase);
    `uvm_info("PHASE", "main phase entered", UVM_LOW)
    phase.raise_objection(this);
    phase.phase_done.set_drain_time(this, 20us);
    #10us;
    phase.drop_objection(this);
    `uvm_info("PHASE", "main phase exited" , UVM_LOW)
```

```
      endtask
      function void report_phase(uvm_phase phase);
        `uvm_info("REPORT", "Test finished", UVM_LOW)
      endfunction
    endclass
  endpackage

  module tb;
    import uvm_pkg::*;
    `include "uvm_macros.svh"
    import rkv_pkg::*;
    initial run_test("drain_time_test");
  endmodule
```

仿真结果：

```
      UVM_INFO @ 0: reporter [RNTST] Running test drain_time_test...
      UVM_INFO uvm_test_drain_time.sv(11) @ 0: uvm_test_top.c1 [PHASE] reset
phase entered
      UVM_INFO uvm_test_drain_time.sv(42) @ 0: uvm_test_top [PHASE] reset phase
entered
      UVM_INFO  uvm_test_drain_time.sv(47)  @ 10000000:  uvm_test_top [PHASE]
reset phase exited
      UVM_INFO uvm_test_drain_time.sv(16) @ 30000000: uvm_test_top.c1 [PHASE]
reset phase exited
      UVM_INFO uvm_test_drain_time.sv(20) @ 90000000: uvm_test_top.c1 [PHASE]
main phase entered
      UVM_INFO uvm_test_drain_time.sv(51) @ 90000000: uvm_test_top [PHASE] main
phase entered
      UVM_INFO  uvm_test_drain_time.sv(56)  @ 100000000:  uvm_test_top [PHASE]
main phase exited
      UVM_INFO uvm_test_drain_time.sv(25) @ 120000000: uvm_test_top.c1 [PHASE]
main phase exited
      UVM_INFO uvm_test_drain_time.sv(60) @ 180000000: uvm_test_top [REPORT]
Test finished
```

阅读手记：

3.1.13 组件的 phase 方法中继承父类的 phase 方法是在做什么？

在大多数情况下，如果是实现用户自定义的子类，那么在定义其各个 phase 方法时，总会默认添加 *super.XXX_phase(phase)* 这样的语句，因为我们清楚这是使子类继承父类的方法。但很多时候我们并不清楚，为什么在继承 uvm_driver、uvm_agent、uvm_sequencer 这些类时调用 *super.XXX_phase(phase)*。

其实，严格来讲，并不是继承每一个 UVM 组件都需要调用父类的 phase 方法，但为简单起见，我们要求遵循这样的规则。例如，继承 uvm_agent 时在 *build_phase()* 调用父类方法，是为了获取上层对它的 is_active 成员的配置，当然，这一步骤也可以显式调用 *uvm_config_db#(T)::get()* 来完成。又例如，继承 uvm_sequencer 时，在其 *connect_phase()* 中调用父类方法，是为了实现内部 fifo 到 sequencer 的 TLM 端口之间的连接，在其 *run_phase()* 中调用父类方法，也是为了能够自动调用上层给指定的 default sequence。

关键词：
phase，super，inheritance

避坑指南：
也许你根本没有时间关心，到底哪些 UVM 组件类在它们各自的 phase 方法中做了什么事情，但是至少应清楚地知道，如果遵循这一简单规则，就会少遇到一些麻烦。此外，对于一些 UVM 组件类，目前很多 phase 方法并没有实现具体的事务，但这并不代表在下一个版本中不会发生变化。所以，这看起来有点多余的一行，还是请习惯性地添加上吧！

阅读手记：

3.1.14 如何控制 UVM 最后打印的消息格式？

如果没有特别指定 report server，我们会选择使用默认的 default report server，它可以通过 *uvm_report_server rps = uvm_report_server::get_server()* 来获取。如果我们对 "Report counts by severity" 的信息之外的其他信息格式如 "Report counts by id" 不感兴趣（它会列举出所有按照消息 ID 做的总结），那么可以进行以下设置来关闭：

```
function void end_of_elaboration_phase(uvm_phase phase);
    uvm_report_server rps = uvm_report_server::get_server();
    uvm_default_report_server drps;
    if(!$cast(drps, rps)) `uvm_error("CASTFAIL", "TYPE CASTING ERROR")
    drps.enable_report_id_count_summary = 0;
endfunction
```

此外，估计很少有人会关闭 UVM 最后按照 severity 级别划分的信息打印吧？例如
** Report counts by severity
UVM_INFO :   13
UVM_WARNING :   0
UVM_ERROR :   2
UVM_FATAL :   0

关键词：
report server，severity，id

避坑指南：
如果还想调整 UVM 最后打印的消息格式，你需要考虑实现自己的 report server。比较容易实现的办法是继承 uvm_default_report_server，并修改它的 *report_summarize()* 函数，同时在顶层利用 *uvm_coreservice_t::set_report_server()* 替代默认的 uvm_default_report_server。

阅读手记：

3.1.15 配置对象的层次为什么应与验证环境的层次相同？

复杂的验证环境往往具备多个验证 VIP，或子一级的验证环境。我们要在验证环境构建和运行前，对这些不同的目标组件进行配置。配置对象（config object）之间的层次保持与验证环境层次一致，以便将它们映射到验证环境一侧的结构中。这种方式有助于理解对验证环境的配置，也有利于自顶向下的例化和配置。

关键词：
configure，environment，component，build，hierarchy，config object

避坑指南：
建议构建与验证环境一致的配置对象层次。不应该直接修改 VIP 内部的属性，而应该利用它提供的配置对象中的属性进行修改，再由顶层进行传递。在传递配置对象时，也建议只传递当前验证环境对应的配置对象,而该配置对象子一级的配置传递应由子一级环境来完成。这种传递方式有助于验证环境的垂直复用（vertical reuse）。

阅读手记：

3.1.16 uvm_config_db 和 uvm_resource_db 在使用时有什么区别？

uvm_resource_db 可以通过静态方法 *set()* 和 *read_by_name()* 完成对某个 scope+name 变量的配置和读取，但它与验证环境层次没有关系。实际上，通过 *uvm_resource_db::set()* 设定的变量读取，默认将遵循 first-write-wins 的规则，因为内部存放的变量是通过队列的形式来实现的，在没有通过 set_override() 设定变量的情况下，在运行过程中对相同 scope+name 变量的第一次配置将会被优先读取。

uvm_config_db 可以通过静态方法 set() 和 get() 完成对某个层次中变量的配置和读取，它遵循的是首先高层次配置覆盖低层次配置，其次是同层次中后面的配置覆盖前面的配置，即 (parent&last)-write-wins。

关键词：

uvm_resource_db，uvm_config_db，set，get

避坑指南：

如果没有特别的需求，uvm_config_db::{set, get} 方法可以满足验证环境变量配置的几乎所有需求，而由于 uvm_resource_db 缺少对验证结构层次的支持，所以应当尽量避免使用 uvm_resource_db。

参考代码： uvm_resource_db_set_get.sv

```
module tb;
  import uvm_pkg::*;
  `include "uvm_macros.svh"
  initial begin
    int count;
    uvm_resource_db#(int)::set("tb","count",10);
    uvm_resource_db#(int)::set("tb","count",20);
    uvm_resource_db#(int)::read_by_name("tb","count",count);
    $display("set-read:: count is %0d", count);
    uvm_resource_db#(int)::set_override("tb","count",30);
    uvm_resource_db#(int)::read_by_name("tb","count",count);
    $display("set_override-read:: count is %0d", count);
  end
endmodule
```

仿真结果：

```
set-read:: count is 10
set_override-read:: count is 30
```

阅读手记：

3.1.17 在继承 UVM 某些参数类时是否需要指定参数类型？

UVM 中有较多的参数类，它们的这种形式（带有参数 type T）也是为了实现更好的复用关系。在继承这些参数类时，可以使用父类默认的参数类型，也可以由子类给定参数类型。

对于独立的参数类（即它的参数类型给定，不需要其他参数类的适配），需要考虑在指定某些参数类型后，这些参数类会影响它的哪些成员变量和成员方法。

对于若干相关联的参数类，如 *uvm_sequencer #(type REQ=uvm_sequence_item, RSP=REQ)* 和 *uvm_driver #(type REQ=uvm_sequence_item, type RSP=REQ)* 这两个参数类，它们的参数类型 *type REQ* 和 *type RSP* 之间就需要保持一致，否则会影响它们之间的 TLM 端口连接。

如果参数类型是父类，那么在子类继承过程中指定的参数类型可以保持不变，或是原参数类型的子类，这都有助于与参数类型有关的父类句柄和子类句柄在代码中的顺利转换。

关键词：
parameterized class，type，TLM，superclass，subclass

避坑指南：
如果要继承参数类，除了可以在继承时指定参数类型，也可以继续保持参数类型的传递链条，这有助于将参数类的形式继承到子类，例如，*class uvm_sequencer #(type REQ=uvm_sequence_item, RSP=REQ) extends uvm_sequencer_param_base #(REQ, RSP)* 这样的子类继承方式。

阅读手记：

3.1.18 UVM 中的注册类有重名时会发生什么？

一般情况下，在一个 package 中不会出现这种问题（同一个 package 不支持同名的类），但是可能在多个 package 中有同名的 UVM 注册类。如果出现这种情况，在仿真运行时 factory（工厂）注册的过程中，只能注册其中的一个类，而另外的类将不能完整地参与类的注册。

尽管接下来的对象例化并不会受到影响，但它仍然会给事后可能应用的类型覆盖（type override）带来问题。因为按照类型名（string 或 uvm_object_wrapper）覆盖，可能会影响到其他相同类名的例化。所以，应尽量避免在 UVM 中出现相同的注册类名。

关键词：
register，class，type，create，override

避坑指南：
如果无法对重名的 UVM 类型进行重命名，那么要注意在以后覆盖类型时，应选择调用函数 *uvm_component::{set_inst_override, set_inst_override_by_type}*，以此指定要覆盖的实例，避免对其他同名类的创建带来影响。

如果只是想移除因为注册的类型名称（string）而带来的重复的 UVM_WARNING 信息，也可以考虑使用宏 *`uvm_component_param_utils* 注册，它只会按照类型注册，所以不会在 factory（工厂）中出现注册后的名称冲突。不过，这种方式也就不再适合通过类型名（string）

来做类型覆盖的操作了。

参考代码：uvm_component_registry_conflicts.sv

```
package rkv_pkg1;
  import uvm_pkg::*;
  `include "uvm_macros.svh"
  class comp1 extends uvm_component;
    `uvm_component_utils(comp1)
    // register only with type but no name
    //`uvm_component_param_utils(comp1)
    function new(string name = "comp1", uvm_component parent = null);
      super.new(name, parent);
    endfunction
  endclass
  class comp1x extends comp1;
    `uvm_component_utils(comp1x)
    function new(string name = "comp1x", uvm_component parent = null);
      super.new(name, parent);
    endfunction
  endclass
  class comp1y extends comp1;
    `uvm_component_utils(comp1y)
    function new(string name = "comp1y", uvm_component parent = null);
      super.new(name, parent);
    endfunction
  endclass
endpackage

package rkv_pkg2;
  import uvm_pkg::*;
  `include "uvm_macros.svh"
  class comp1 extends uvm_component;
    `uvm_component_utils(comp1)
    // register only with type but no name
    //`uvm_component_param_utils(comp1)
    function new(string name = "comp1", uvm_component parent = null);
      super.new(name, parent);
    endfunction
  endclass
  class comp1z extends comp1;
    `uvm_component_utils(comp1z)
    function new(string name = "comp1z", uvm_component parent = null);
      super.new(name, parent);
    endfunction
  endclass
endpackage
```

```
package rkv_pkg3;
  import uvm_pkg::*;
  `include "uvm_macros.svh"
  class component_registry_conflicts_test extends uvm_test;
    rkv_pkg1::comp1 c0;
    rkv_pkg2::comp1 c1;
    `uvm_component_utils(component_registry_conflicts_test)
    function new(string name = "component_registry_conflicts_test",
                 uvm_component parent = null);
      super.new(name, parent);
    endfunction
    function void build_phase(uvm_phase phase);
      super.build_phase(phase);
      // ERROR override would effect irrelated types
      //set_type_override("comp1", "comp1x");
      //set_type_override_by_type(rkv_pkg1::comp1::get_type(),
      //                          rkv_pkg1::comp1y::get_type());
      //set_type_override_by_type(rkv_pkg2::comp1::get_type(),
      //                          rkv_pkg2::comp1z::get_type());

      // Available to specify the overrided instance than type
      //set_inst_override("c0", "comp1", "comp1x");
      //set_inst_override_by_type("c1", rkv_pkg2::comp1::get_type(),
                                 rkv_pkg2::comp1z::get_type());

      c0 = rkv_pkg1::comp1::type_id::create("c0", this);
      c1 = rkv_pkg2::comp1::type_id::create("c1", this);
      `uvm_info("build", "exited", UVM_LOW)
    endfunction
  endclass
endpackage
```

仿真结果：

UVM_WARNING @ 0: reporter [TPRGED] Type name 'comp1' already registered with factory. No string-based lookup support for multiple types with the same type name.

UVM_INFO @ 0: reporter [RNTST] Running test component_registry_conflicts_test...

Hierarchy	Class Type	Object ID	Create Time
top_levels	uvm_component[]		
[0]	component_registry_conflicts_test	@1	0
c0	comp1	@1	0
c1	comp1	@1	0

comp1 类型组件被替换后的实例信息

参考代码：uvm_factory.svh（UVM 源代码）

```
function void uvm_default_factory::register (uvm_object_wrapper obj);
  if (obj == null) begin
    uvm_report_fatal ("NULLWR",
      "Attempting to register a null object with the factory", UVM_NONE);
  end
  if (obj.get_type_name() != "" && obj.get_type_name() != "<unknown>")
  begin
    if (m_type_names.exists(obj.get_type_name()))
      uvm_report_warning("TPRGED", {"Type name '",obj.get_type_name(),
        "' already registered with factory. No string-based lookup ",
        "support for multiple types with the same type name."}, UVM_NONE);
    else
      m_type_names[obj.get_type_name()] = obj;
  end

  if (m_types.exists(obj)) begin
    if (obj.get_type_name() != "" && obj.get_type_name() != "<unknown>")
      uvm_report_warning("TPRGED", {"Object type '",obj.get_type_name(),
                   "' already registered with factory. "}, UVM_NONE);
  end
  ...
endfunction
```

阅读手记：

3.1.19 为什么在 build_phase 中访问更低层次的组件会失败呢？

在 build_phase 中访问更低层次的组件或其成员出现失败，是由于底层组件此时并未创建好，或者底层组件的成员还未准备好供外部调用。需要理解的是，上层组件虽然通过 factory 创建了子一级组件，但子一级组件只是被创建了（间接调用 new() 函数），并未执行 build_phase() 函数。这时，只能看做子一级组件被例化了，但子一级组件的 build_phase 并未执行（需要等到上一级组件 build_phase() 执行完，才能轮到子一级组件执行其 build_phase() 函数）。当然，比子一级组件更低层次的组件在此时也就无法访问了。这一执行顺序，使得用户在访问一些低层次的组件或其成员时会发生运行时错误（如目标句柄为空）。

关键词：
build_phase，new，create，component，handle，null

避坑指南：

建议在整个结构都准备好后再访问更底层次的组件及其成员。这个建议的 phase 可以是 *connect_phase()* 或 *end_of_elaboration_phase()*。在一般情况下，我们建议通过句柄访问下一级组件或其成员。而对于跨层次较多的访问，我们并不建议这么做，因为这会增加组件之间的耦合性，而且会让验证环境越来越难以维护和阅读。

参考代码： uvm_lower_hier_access_issue.sv

```
package rkv_pkg;
  import uvm_pkg::*;
  `include "uvm_macros.svh"
  class rkv_config extends uvm_object;
    bit ctrl0;
    bit ctrl1;
    `uvm_object_utils(rkv_config)
    function new(string name = "rkv_config");
      super.new(name);
    endfunction
  endclass
  class rkv_component extends uvm_component;
    bit ctrl0;
    bit ctrl1;
    rkv_config cfg;
    `uvm_component_utils(rkv_component)
    function new(string name = "rkv_component",
                 uvm_component parent = null);
      super.new(name, parent);
    endfunction
    function void build_phase(uvm_phase phase);
      if(!uvm_config_db#(rkv_config)::get(this,"","cfg", cfg)) begin
        `uvm_error("GETCFG","cannot get rkv_config handle from config DB")
      end
    endfunction
  endclass
  class rkv_env extends uvm_component;
    rkv_component comp;
    `uvm_component_utils(rkv_env)
    function new(string name = "rkv_env", uvm_component parent = null);
      super.new(name, parent);
    endfunction
    function void build_phase(uvm_phase phase);
      comp = rkv_component::type_id::create("comp", this);
      // available access
      comp.ctrl0 = 0;
      // ERROR access
      // comp.cfg.ctrl0 = 1;
    endfunction
```

```systemverilog
    function void end_of_elaboration_phase(uvm_phase phase);
      // available access
      comp.ctrl0 = 0;
      comp.cfg.ctrl0 = 1;
    endfunction
  endclass
  class lower_hier_access_issue_test extends uvm_component;
    rkv_config cfg;
    rkv_env env;
    `uvm_component_utils(lower_hier_access_issue_test)
    function new(string name = "lower_hier_access_issue_test",
                 uvm_component parent = null);
      super.new(name, parent);
    endfunction
    function void build_phase(uvm_phase phase);
      cfg = rkv_config::type_id::create("cfg", this);
      env = rkv_env::type_id::create("env", this);
      uvm_config_db#(rkv_config)::set(this,"env.comp","cfg", cfg);
      // ERROR access
      // env.comp.ctrl1 = 1;
      // env.comp.cfg.ctrl1 = 1;
    endfunction
    function void end_of_elaboration_phase(uvm_phase phase);
      // available access
      env.comp.ctrl1 = 1;
      env.comp.cfg.ctrl1 = 1;
    endfunction
  endclass
endpackage

module tb;
  import uvm_pkg::*;
  `include "uvm_macros.svh"
  import rkv_pkg::*;
  initial run_test("lower_hier_access_issue_test");
endmodule
```

阅读手记：

3.1.20 在 UVM 中是否可跨层次调用某些组件的方法？

如果要跨层次去调用某些组件的方法，就无异于抛弃 TLM 端口连接的方式，而在 initiator 一侧直接访问 target 处的成员（通过跨层次获得 target 的句柄）。我们可以理解的是，TLM 端口和连接方式存在的意义就是做好组件和环境之间的独立性，减少组件之间的耦合性。

我们建议减少跨层次去调用某些组件的方法。更准确地说，应该避免在独立组件（uvm_component）之间、独立环境（uvm_env）之间进行跨层次访问。如果要访问，那么调用发生的地方只应该是在更上层的 test 和 sequence 处。这是因为 test 和 sequence 是伴随测试场景的，它们在环境、组件复用移植的过程中参与得不多。此外，若有必要，也可以将部分接口通过层层传递的方式抵达更上层，以便将需要的低层次接口暴露出来。

关键词：
hierarchy，cross-hierarchy，call，component，environment，test，sequence

避坑指南：
如果要去跨层次调用某些组件的方法，还要注意的是，在不同的环境结构配置下，这些跨层次的组件可能不存在，那么就还是要做好更多的场景适配。

参考代码： uvm_cross_hier_access.sv

```
package rkv_pkg;
  import uvm_pkg::*;
  `include "uvm_macros.svh"
  class rkv_config extends uvm_object;
    int comp_num = 1;
    `uvm_object_utils(rkv_config)
    function new(string name = "rkv_config");
      super.new(name);
    endfunction
  endclass
  class rkv_component extends uvm_component;
    local bit _ctrl = 1;
    `uvm_component_utils(rkv_component)
    function new(string name = "rkv_component",
                 uvm_component parent = null);
      super.new(name, parent);
    endfunction
    function bit get_ctrl();
      return _ctrl;
    endfunction
  endclass
  class rkv_env extends uvm_component;
    rkv_config cfg;
    rkv_component comps[];
    `uvm_component_utils(rkv_env)
    function new(string name = "rkv_env", uvm_component parent = null);
      super.new(name, parent);
    endfunction
    function void build_phase(uvm_phase phase);
      if(!uvm_config_db#(rkv_config)::get(this,"","cfg", cfg)) begin
        `uvm_error("GETCFG","cannot get rkv_config handle from config DB")
      end
      comps = new[cfg.comp_num];
      for(int i=0; i<cfg.comp_num; i++)
```

```
            comps[i] = rkv_component::type_id::create(
                       $sformatf("comps[%0d]", i), this);
        endfunction
        function bit get_ctrl(int id = 0);
          if(id < cfg.comp_num && comps[id] != null)
            return comps[id].get_ctrl();
          else
            return 0;
        endfunction
      endclass
      class cross_hier_access_test extends uvm_component;
        rkv_config cfg;
        rkv_env env;
        `uvm_component_utils(cross_hier_access_test)
        function new(string name = "cross_hier_access_test",
                    uvm_component parent = null);
          super.new(name, parent);
        endfunction
        function void build_phase(uvm_phase phase);
          cfg = rkv_config::type_id::create("cfg", this);
          cfg.comp_num = 3;
          uvm_config_db#(rkv_config)::set(this,"env","cfg", cfg);
          env = rkv_env::type_id::create("env", this);
        endfunction
        task run_phase(uvm_phase phase);
          bit ctrl;
          phase.raise_objection(this);
          // suggested way
          ctrl = env.get_ctrl(2);
          `uvm_info("GETVAL",
                   $sformatf("comp[2] ctrl bit is %0d", ctrl), UVM_LOW)
          // available but not best
          ctrl = env.comps[1].get_ctrl();
          `uvm_info("GETVAL",
                   $sformatf("comp[1] ctrl bit is %0d", ctrl), UVM_LOW)
          phase.drop_objection(this);
        endtask
      endclass
    endpackage

    module tb;
      import uvm_pkg::*;
      `include "uvm_macros.svh"
      import rkv_pkg::*;
      initial run_test("cross_hier_access_test");
    endmodule
```

仿真结果：

```
    UVM_INFO @ 0: reporter [RNTST] Running test cross_hier_access_test...
    UVM_INFO uvm_cross_hier_access.sv(61) @ 0: uvm_test_top [GETVAL] comp[2]
ctrl bit is 1
```

```
        UVM_INFO uvm_cross_hier_access.sv(64) @ 0: uvm_test_top [GETVAL] comp[1]
ctrl bit is 1
```

阅读手记：

3.1.21 在使用 uvm_config_db 时要注意什么？

在使用 uvm_config_db 进行配置（set）和获取（get）的操作时，需要理解 set 的路径（具体路径或包含通配符）与 get 的路径是否能够有共同的地方（即可以 get 到配置数据），以及 set 的类型与 get 的类型是否严格一致。

关键词：

uvm_config_db，set，get，path，parameterized，type

避坑指南：

执行 *uvm_config_db#(T)::get()* 操作时所在的对象路径可以与传入参数的组件路径不一致，只要传入参数的路径能够获得目标类型的数据即可。如果 *uvm_config_db#(T)* 中的参数类型自身也为参数化类型（例如，参数化的类、参数化的接口），那么 *uvm_config_db#(T)::{set, get}* 双方所用的参数化类型也必须保持一致。

参考代码： uvm_config_db_set_get_matched.sv

```
package rkv_pkg;
  import uvm_pkg::*;
  `include "uvm_macros.svh"
  class rkv_config #(type T = int) extends uvm_object;
    T data[];
    `uvm_object_param_utils(rkv_config)
    function new(string name = "rkv_config");
      super.new(name);
    endfunction
  endclass
  class rkv_component extends uvm_component;
    int ctrl0;
    byte ctrl1;
    bit ctrl2;
    rkv_config#(byte unsigned) cfg;
    `uvm_component_utils(rkv_component)
    function new(string name = "rkv_component",
                 uvm_component parent = null);
      super.new(name, parent);
    endfunction
    function void build_phase(uvm_phase phase);
```

```systemverilog
      if(!uvm_config_db#(rkv_config#(byte unsigned))::get(
          get_parent(),"","cfg", cfg))
        begin
          `uvm_warning("GETCFG",
            "cannot get rkv_config #(byte unsigned) handle from config DB")
        end
        if(!uvm_config_db#(int)::get(this,"","ctrl", ctrl0)) begin
          `uvm_warning("GETCFG","cannot get ctrl (int type) from config DB")
        end
        if(!uvm_config_db#(byte)::get(this,"","ctrl", ctrl1)) begin
          `uvm_warning("GETCFG","cannot get ctrl (byte type) from config DB")
        end
        if(!uvm_config_db#(bit)::get(this,"","ctrl", ctrl2)) begin
          `uvm_warning("GETCFG","cannot get ctrl (bit type) from config DB")
        end
      endfunction
    endclass
    class rkv_env extends uvm_component;
      uvm_object cfg0;
      rkv_config cfg1;
      rkv_config#(byte) cfg2;
      rkv_config#(byte unsigned) cfg3;
      rkv_component comp;
      `uvm_component_utils(rkv_env)
      function new(string name = "rkv_env", uvm_component parent = null);
        super.new(name, parent);
      endfunction
      function void build_phase(uvm_phase phase);
        if(!uvm_config_db#(uvm_object)::get(this,"","cfg", cfg0)) begin
          `uvm_warning("GETCFG",
                    "cannot get uvm_object handle from config DB")
        end
        if(!uvm_config_db#(rkv_config)::get(this,"","cfg", cfg1)) begin
          `uvm_warning("GETCFG",
                    "cannot get rkv_config #(int) handle from config DB")
        end
        if(!uvm_config_db#(rkv_config#(byte))::get(this,"","cfg", cfg2))
         begin
          `uvm_warning("GETCFG",
                    "cannot get rkv_config #(byte) handle from config DB")
        end
        if(!uvm_config_db#(rkv_config#(byte unsigned))::get(
            this,"","cfg", cfg3)) begin
          `uvm_warning("GETCFG",
            "cannot get rkv_config #(byte unsigned) handle from config DB")
        end
        comp = rkv_component::type_id::create("comp", this);
      endfunction
    endclass
    class config_set_get_matched_test extends uvm_component;
      rkv_config #(byte unsigned) cfg;
```

```
    rkv_env env;
    `uvm_component_utils(config_set_get_matched_test)
    function new(string name = "config_set_get_matched_test",
                 uvm_component parent = null);
      super.new(name, parent);
    endfunction
    function void build_phase(uvm_phase phase);
      cfg = new("cfg");
      //cfg = rkv_config#(byte unsigned)::type_id::create("cfg", this);
      uvm_config_db#(rkv_config#(byte unsigned))::set(
        this,"env","cfg", cfg);
      uvm_config_db#(bit)::set(this,"env*","ctrl", 1);
      env = rkv_env::type_id::create("env", this);
    endfunction
    task run_phase(uvm_phase phase);
      phase.raise_objection(this);
      phase.drop_objection(this);
    endtask
  endclass
endpackage

module tb;
  import uvm_pkg::*;
  `include "uvm_macros.svh"
  import rkv_pkg::*;
  initial run_test("config_set_get_matched_test");
endmodule
```

仿真结果：

```
    UVM_INFO @ 0: reporter [RNTST] Running test config_set_get_matched_test...
    UVM_WARNING uvm_config_db_set_get_matched.sv(47) @ 0: uvm_test_top.env
[GETCFG] cannot get uvm_object handle from config DB
    UVM_WARNING uvm_config_db_set_get_matched.sv(50) @ 0: uvm_test_top.env
[GETCFG] cannot get rkv_config #(int) handle from config DB
    UVM_WARNING uvm_config_db_set_get_matched.sv(53) @ 0: uvm_test_top.env
[GETCFG] cannot get rkv_config #(byte) handle from config DB
    UVM_WARNING uvm_config_db_set_get_matched.sv(25) @ 0: uvm_test_top.env.
comp [GETCFG] cannot get ctrl (int type) from config DB
    UVM_WARNING uvm_config_db_set_get_matched.sv(28) @ 0: uvm_test_top.env.
comp [GETCFG] cannot get ctrl (byte type) from config DB
```

阅读手记：

3.2 通信与同步

SystemVerilog 对通信与同步也有支持，而 UVM 更侧重于在验证组件之间进行规范的通信方法与同步方法。掌握 TLM 通信端口和方法实现之后，用户需要恰当地在组件中创建 TLM FIFO。对于偶发性的事件触发、同步和数据传递，需要更好地使用 uvm_event。预置事件和回调函数的 UVM 类型有助于类的继承和复用。接下来我们将围绕 UVM 的通信和同步讨论若干疑难点。

3.2.1 event 和 uvm_event 的联系和区别是什么？

SystemVerilog 的 event 虽然不需要例化（*new()*），但其使用起来也与 uvm_event 类似，可以通过@或 *event::triggered()* 来等待事件。uvm_event 需要例化，可以通过 *wait_trigger()*、*wait_ptrigger()* 来等待事件。uvm_event 还可以在触发时传递数据，并通过 *wait_trigger_data()*、*wait_ptrigger_data()* 在等到事件的同时获取数据。

关键词：
event，uvm_event，trigger

避坑指南：
电平方式的等待触发可选择 *event::triggered()* 或 *uvm_event::wait_ptrigger()*。如果事件仅是为了在类或某个域局部使用，那么 event 可以满足需要。如果事件不仅是在当前类或域中使用，还可能被其他类或域所需要，那么 uvm_event 与 uvm_event_pool 一起使用更为合适。

阅读手记：

3.2.2 sequencer 和 driver 的类型参数 REQ/RSP 需要保持一致吗？

是的，它们的参数类型 REQ、RSP 需要严格保持一致。因为这两个类的参数决定了接下来的 TLM 端口类型和连接要求。在 TLM 通信中，要求 port、export、imp 在连接时，传输的类型必须保持一致（也不允许将父类和子类混用为传输类型参数）。

关键词：
sequencer，driver，type，REQ，RSP

避坑指南：

如果 sequencer 和 driver 在定义时采用了默认参数 uvm_sequence_item，那么在 *driver::get_next_item(REQ)* 获得 REQ 之后，往往需要将 REQ 类型（父类句柄）转换为子类句柄，这样才能访问其子类对象的成员。

阅读手记：

3.2.3 TLM FIFO 的方法是否可以直接调用？

当然可以直接调用 uvm_tlm_fifo、uvm_tlm_analysis_fifo 的方法。无论是从例化的 PORT 端口（如 uvm_blocking_put_port）调用 TLM FIFO 一侧的方法（例如 *put()*），还是在例化的 TLM FIFO 一侧的组件直接调用其方法（如 *get()*），都是可以的。

关键词：

uvm_tlm_fifo，uvm_tlm_analysis_fifo，TLM，port

避坑指南：

由于 TLM FIFO 已经预置了数据缓存（mailbox 实现）和方法（如 *put()*、*get()*、*write()*），所以，无论 PORT 一侧是在连接 TLM FIFO 后调用其方法，还是通过 TLM FIFO 句柄调用其方法，都是可以的。

参考代码： uvm_tlm_fifo_put_get.sv

```
package rkv_pkg;
  import uvm_pkg::*;
  `include "uvm_macros.svh"

  class rkv_trans extends uvm_object;
    int data;
    `uvm_object_utils(rkv_trans)
    function new(string name = "rkv_trans");
      super.new(name);
    endfunction
  endclass

  class rkv_fifo #(type T = rkv_trans) extends uvm_tlm_analysis_fifo #(T);
    int buffer [];
    function new(string name = "rkv_fifo", uvm_component parent = null);
      super.new(name, parent);
    endfunction
    task get_data(int ntrans);
```

```
      T t;
      repeat(ntrans) begin
        get(t); // blocking get data (data consuming behaivor)
        buffer = {buffer, t.data}; // compose a new dynamic array
         `uvm_info("GETDATA",
           $sformatf("@time %0t::buffer content is %p", $time, buffer),
           UVM_LOW)
      end
    endtask
  endclass

  class tlm_fifo_put_get_test extends uvm_test;
    rkv_fifo fifo;
     `uvm_component_utils(tlm_fifo_put_get_test)
    function new(string name = "tlm_fifo_put_get_test",
                 uvm_component parent = null);
      super.new(name, parent);
    endfunction
    function void build_phase(uvm_phase phase);
      super.build_phase(phase);
      fifo = new("fifo", this);
    endfunction

    task run_phase(uvm_phase phase);
      rkv_trans t;
      int ntr = 5;
      super.run_phase(phase);
      phase.raise_objection(this);
      fork
        // mimic the data pushed into the target analysis fifo
        begin
          for(int i=1; i<=ntr; i++) begin
            t = new();
            t.data = i;
            fifo.put(t);
            #10ns;
          end
        end
        // mimic the data popped from fifo in parallel
        begin
          fifo.get_data(ntr);
        end
      join
      phase.drop_objection(this);
    endtask
  endclass
endpackage
```

```
module tb;
  import uvm_pkg::*;
  `include "uvm_macros.svh"
  import rkv_pkg::*;
  initial run_test("tlm_fifo_put_get_test");
endmodule
```

仿真结果：

```
    UVM_INFO @ 0: reporter [RNTST] Running test tlm_fifo_put_get_test...
    UVM_INFO uvm_tlm_fifo_put_get.sv(23) @ 0: uvm_test_top.fifo [GETDATA]
@time 0::buffer content is '{1}
    UVM_INFO uvm_tlm_fifo_put_get.sv(23) @ 10000: uvm_test_top.fifo [GETDATA]
@time 10000::buffer content is '{1, 2}
    UVM_INFO uvm_tlm_fifo_put_get.sv(23) @ 20000: uvm_test_top.fifo [GETDATA]
@time 20000::buffer content is '{1, 2, 3}
    UVM_INFO uvm_tlm_fifo_put_get.sv(23) @ 30000: uvm_test_top.fifo [GETDATA]
@time 30000::buffer content is '{1, 2, 3, 4}
    UVM_INFO uvm_tlm_fifo_put_get.sv(23) @ 40000: uvm_test_top.fifo [GETDATA]
@time 40000::buffer content is '{1, 2, 3, 4, 5}
```

阅读手记：

3.2.4 为什么 uvm_object 类型不能例化 TLM 端口？

首先，无论是 TLM 端口在例化时调用 new() 函数时需要的组件类型参数要求，还是 TLM imp 端口在声明时需要绑定的组件类型的要求（`uvm_xxx_imp #(imp, TYPE, arg)`），都规定了 TLM 端口只应该在组件中使用和例化。

其次，从应用层面而言，uvm_object 这样可能随时被销毁的对象，并没有在仿真过程中进行持续通信的需求，即便有，也可以通过其他方式实现。但是，如果采用 TLM 端口，那么在通信的过程中甚至无法保证目标端（target）长期存在。目标端一旦被销毁，接下来的通信可能将无法进行。

关键词：
uvm_object，TLM，port

避坑指南：
如果是 uvm_sequence（uvm_object 类型）与验证环境中的组件通信，可以通过顶层传递的 config object 或 virtual interface 中的变量完成等待和触发，也可以通过 *uvm_event:: trigger(T data)* 完成事件的触发和数据的发送。如果是 uvm_sequence 之间的通信，那么建议利用顶层

virtual sequence 实现整个测试场景的调度协调。

阅读手记：

3.2.5 UVM 回调函数类的使用特点有哪些？

不管是 UVM 回调函数还是 SystemVerilog 回调函数，都有以下几个特点。第一，需要定义回调函数类和对应的回调方法；第二，在目标类的某些方法中调用这些回调函数类的方法（支持顶层绑定多个回调函数类）；第三，在顶层例化回调函数对象，再将这些回调函数对象与目标对象进行绑定。通过这三个步骤就能够完成 UVM 回调函数类从定义到嵌入、再到绑定的整个过程。

关键词：
uvm_callback，method，bind

避坑指南：
要完成以上三个步骤，需要将这些类、宏、方法正确嵌入到过程中：*uvm_callback*，`*uvm_register_cb(T, CB)*，`*uvm_do_callbacks(T, CB, METHOD)*，*uvm_callbacks #(T, CB)::add(obj, cb)*。

阅读手记：

3.2.6 uvm_event 应从哪里例化和获取？

在 UVM 环境的各个组件之间，除了可以通过 TLM 端口完成常规的数据传输，对于一些偶发性的事件，也可以通过 uvm_event 来完成广播通知。uvm_event 的触发并不局限于某个 uvm_component 内，它的触发和接收也可以在诸如 sequence 这类 uvm_object 类型之中进行。所以从使用范围看，对于偶然触发的广播行为，uvm_event 可以更好地胜任。

可以创建并获得 uvm_event 句柄的简单方法是 *uvm_event::get_global()*。这一方法通过全局唯一的 event global pool 根据名称来创建 uvm_event 实例。在实例创建好之后，创建 uvm_event 的实例以及其他 UVM 组件都可以通过该 event 来完成同步。

此外，也可以通过 uvm local pool 的方式完成 uvm_event 实例的创建，但如果组件之间共用这个 uvm_event，则只能通过层次化句柄引用来完成。如果要通过 global pool 创建 uvm_event

实例，则需考虑的是，通过 *uvm_event_pool::get_global()* 给定相同 uvm_event 名称时，它们都将获得同一个 uvm_event 实例。这对于可能例化多次的组件不见得是一个完美的解决办法。而在通过 global pool 创建 uvm_event 实例时传递独一无二的名称，或通过 local pool 创建 uvm_event，都能够获得独立创建的实例，避免 uvm_component 组件在例化多次后，仍然只能够获得相同 uvm_event 实例的句柄。

关键词：
uvm_event，uvm_event_pool，global，local

避坑指南：
这里并非在简单地推荐使用 global pool 或 local pool，而是提醒用户自己考虑应该在哪些情况下使用 global pool、在哪些情况下使用 local pool，以及在调用 *uvm_event_pool::get_global()* 时是应该传递相同的名称还是传递不同的名称，以便获得期望的 uvm_event 实例句柄。这里给的建议是，在 VIP 组件（如 monitor）内部的 uvm_event 创建，采用 local pool；在验证环境组件（如 scoreboard）内部的 uvm_event 创建，采用 global pool。

参考代码： uvm_event_global_vs_local.sv

```
package rkv_pkg;
  import uvm_pkg::*;
  `include "uvm_macros.svh"
  class comp1 extends uvm_component;
    uvm_event glb_e0, glb_e1;
    uvm_event loc_e0;
    `uvm_component_utils(comp1)
    function new(string name = "comp1", uvm_component parent = null);
      super.new(name, parent);
    endfunction
    function void build_phase(uvm_phase phase);
      uvm_event_pool ep = new("ep");
      glb_e0 = uvm_event_pool::get_global("glb_e0");
      glb_e1 = uvm_event_pool::get_global(
                 $sformatf("%s.glb_e1",get_full_name()));
      loc_e0 = ep.get("loc_e0");
    endfunction
  endclass
  class event_global_vs_local_test extends uvm_test;
    comp1 c0, c1;
    `uvm_component_utils(event_global_vs_local_test)
    function new(string name = "event_global_vs_local_test",
                 uvm_component parent = null);
      super.new(name, parent);
    endfunction
    function void build_phase(uvm_phase phase);
      c0 = comp1::type_id::create("c0", this);
      c1 = comp1::type_id::create("c1", this);
```

```
      endfunction
    task run_phase(uvm_phase phase);
      phase.raise_objection(this);
      if(c0.glb_e0 == c1.glb_e0)
        `uvm_info("HDLCMP", "c0.glb_e0 == c1.glb_e0", UVM_LOW)
      if(c0.glb_e1 == c1.glb_e1)
        `uvm_info("HDLCMP", "c0.glb_e1 == c1.glb_e1", UVM_LOW)
      if(c0.loc_e0 == c1.loc_e0)
        `uvm_info("HDLCMP", "c0.loc_e0 == c1.loc_e0", UVM_LOW)
      phase.drop_objection(this);
    endtask
  endclass
endpackage

module tb;
  import uvm_pkg::*;
  `include "uvm_macros.svh"
  import rkv_pkg::*;
  initial run_test("event_global_vs_local_test");
endmodule
```

仿真结果：

```
UVM_INFO @ 0: reporter [RNTST] Running test event_global_vs_local_test...
UVM_INFO uvm_event_global_vs_local.sv(31) @ 0: uvm_test_top [HDLCMP] c0.glb_e0 == c1.glb_e0
```

阅读手记：

3.2.7 TLM 端口为什么没有注册过呢？

在使用 TLM 相关类时，可能对创建 TLM 端口和 TLM FIFO 产生疑问。为什么在创建它们时没有（且无法）使用工厂去创建呢？这是因为，在 UVM 库中的这两个大类并没有做过注册，所以无法使用工厂去创建。我们需要弄清楚的是，工厂的存在是为了便于以后可能发生的类型覆盖，而对于 TLM 端口和 TLM FIFO，它们的功能单一且固定，所以不会对这两种大类使用工厂做覆盖，而且 UVM 库中的多数类也没有做过工厂注册。

关键词：
TLM，port，register，factory，override

避坑指南：
如果基于 TLM FIFO 进行子类扩展，那么可以对其进行注册（如果必要），再使用工厂去创建。但我们无法对扩展的 TLM 端口进行注册，因为它们并非继承于 uvm_object 类型，不满足注册的类型要求。

参考代码：uvm_factory_tlm_fifo_register.sv

```
package rkv_pkg;
  import uvm_pkg::*;
  `include "uvm_macros.svh"
  class user_fifo_str extends uvm_tlm_fifo #(string);
    `uvm_component_utils(user_fifo_str)
    function new(string name = "user_fifo_str",
                 uvm_component parent = null);
      super.new(name, parent);
    endfunction
  endclass

  class tlm_fifo_register_test extends uvm_test;
    user_fifo_str fifo0;
    `uvm_component_utils(tlm_fifo_register_test)
    function new(string name = "tlm_fifo_register_test",
                 uvm_component parent = null);
      super.new(name, parent);
    endfunction
    function void build_phase(uvm_phase phase);
      fifo0 = user_fifo_str::type_id::create("fifo0", this);
    endfunction
  endclass
endpackage

module tb;
  import uvm_pkg::*;
  `include "uvm_macros.svh"
  import rkv_pkg::*;
  initial run_test("tlm_fifo_register_test");
endmodule
```

阅读手记：

3.3 测试序列

待 UVM 的验证结构建立起来之后，就需要更多地考虑测试序列的组织和发送问题。发送测试序列时会遇到序列挂载方式的问题、序列防止退出的问题、安全终止序列的问题、序列获得配置变量的问题、序列监测变量和覆盖率的问题等一系列要考虑的疑难点。用户在项目实际中，将会遇到以下具体的问题，我们对此一一展开讨论。

3.3.1 m_sequencer 和 p_sequencer 有什么区别？

m_sequencer 是 sequence 或 sequence item 的成员变量，即 sequence、sequence item 一旦"挂载"到某个 sequencer 上，那么该 sequencer 的句柄即被赋值给 m_sequencer（uvm_sequencer_base 类）。

p_sequencer 通常需要在定义 sequence 类时，通过宏 `uvm_declare_p_sequencer(SEQUENCER) 进行声明（间接定义 p_sequencer 成员变量）。

m_sequencer 与 p_sequencer 均指向同一个挂载的 sequencer 实例，前者是父类句柄类型，后者是子类句柄类型。

关键词：
m_sequencer，p_sequencer，delcare

避坑指南：
如果 sequence 需要访问所挂载 sequencer 子类的成员，那么必须使用 p_sequencer 子类句柄，这也是 virtual sequence 在定义时经常通过宏来间接声明 p_sequencer 句柄的原因。

参考代码： uvm_sequencer_m_p_reference.sv

```
package rkv_pkg;
  import uvm_pkg::*;
  `include "uvm_macros.svh"
  class rkv_sequencer extends uvm_sequencer;
    bit ctrl;
    bit flag;
    `uvm_component_utils(rkv_sequencer)
    function new(string name = "rkv_sequencer",
                 uvm_component parent = null);
      super.new(name, parent);
    endfunction
  endclass
  class rkv_driver extends uvm_driver;
    `uvm_component_utils(rkv_driver)
    function new(string name = "rkv_driver", uvm_component parent = null);
      super.new(name, parent);
    endfunction
```

```systemverilog
    endclass
    class rkv_sequence extends uvm_sequence;
      `uvm_declare_p_sequencer(rkv_sequencer)
      `uvm_object_utils(rkv_sequence)
      function new(string name = "rkv_sequence");
        super.new(name);
      endfunction
      task body();
        // access sequencer member via m_sequencer
        m_sequencer.grab(this);
        m_sequencer.set_arbitration(UVM_SEQ_ARB_WEIGHTED);
        // access sequencer member via p_sequencer
        p_sequencer.ctrl = 1;
        p_sequencer.flag = 1;
        m_sequencer.ungrab(this);
      endtask
    endclass
    class sequencer_reference_test extends uvm_test;
      rkv_driver driver;
      rkv_sequencer sequencer;
      `uvm_component_utils(sequencer_reference_test)
      function new(string name = "sequencer_reference_test",
                   uvm_component parent = null);
        super.new(name, parent);
      endfunction
      function void build_phase(uvm_phase phase);
        driver = rkv_driver::type_id::create("driver", this);
        sequencer = rkv_sequencer::type_id::create("sequencer", this);
      endfunction
      function void connect_phase(uvm_phase phase);
        driver.seq_item_port.connect(sequencer.seq_item_export);
      endfunction
      task run_phase(uvm_phase phase);
        rkv_sequence seq = rkv_sequence::type_id::create("seq");
        phase.raise_objection(this);
        seq.start(sequencer);
        phase.drop_objection(this);
      endtask
    endclass
endpackage

module tb;
  import uvm_pkg::*;
  `include "uvm_macros.svh"
  import rkv_pkg::*;
  initial run_test("sequencer_reference_test");
endmodule
```

阅读手记：

3.3.2 为什么不建议在 sequence 中使用 pre_body()和 post_body()？

因为只有通过 *uvm_sequence::start()* 方法调用时，才会间接调用 *pre_body()* 和 *post_body()*（*start()* 方法中的参数 call_pre_post 默认值为 1）。而经常使用的序列发送宏如 \`uvm_do、\`uvm_do_on，则不会调用 *pre_body()* 和 *post_body()*（宏内部在调用 *sequence::start()* 方法时，将 call_pre_post 参数设定为 0）。

关键词：

start，uvm_do，uvm_do_on，pre_body，post_body

避坑指南：

为实现 *pre_body()*、*post_body()* 的执行效果，可以定义新任务 *body_pre()* 和 *body_post()* 并且将其置于 *body()* 中。或者，在父类 *body()* 中前置执行任务，在子类 *body()* 中通过 *super.body()* 继承父类的前置任务。

参考代码： uvm_sequence_defines.svh（UVM 源代码）

```
`define uvm_do_on_pri_with(SEQ_OR_ITEM, SEQR, PRIORITY, CONSTRAINTS) \
  begin \
  uvm_sequence_base __seq; \
  `uvm_create_on(SEQ_OR_ITEM, SEQR) \
  if (!$cast(__seq,SEQ_OR_ITEM)) start_item(SEQ_OR_ITEM, PRIORITY);\
  if ((__seq == null || !__seq.do_not_randomize) && \
      !SEQ_OR_ITEM.randomize() with CONSTRAINTS ) begin \
    `uvm_warning("RNDFLD", "Randomization failed in uvm_do_with action") \
  end\
  if (!$cast(__seq,SEQ_OR_ITEM)) finish_item(SEQ_OR_ITEM, PRIORITY); \
  // start 方法原型
  // uvm_sequence::start (uvm_sequencer_base sequencer,
  //                     uvm_sequence_base parent_sequence = null,
  //                     int this_priority = -1,
  //                     bit call_pre_post = 1);)
  else __seq.start(SEQR, this, PRIORITY, 0); \
  end
```

阅读手记：

3.3.3 sequence 如何通过 uvm_config_db 获得配置的变量？

第一种方法，sequence 可以通过其所挂载的 sequencer 和其所在的配置层次，间接获得配置的变量。第二种方法，可以在创建 sequence 时利用 *create()* 函数为其指定 parent component，继而获取 uvm_config_db 配置结构中的层次位置，这样可以利用这个层次位置去配置变量。第三种方法更为直接，只要顶层配置时与 sequence 做好对应配置层次和变量名的约定，sequence 依然可以获得配置的变量。

关键词：
sequence，component，create，uvm_config_db

避坑指南：
sequence 本身不是 uvm_component 类型，无法作为 *uvm_config_db::get(uvm_component cntxt, ...)* 的第一个参数 cntxt，但可以通过 m_sequencer 来获得配置的变量。利用 m_sequencer 作为在验证层次中相对的"位置锚点"，那么通过这个 m_sequencer 句柄来获得变量的方式，更利于后期验证层次变化后的复用。

参考代码： uvm_sequence_get_from_config_db.sv

```
package rkv_pkg;
  import uvm_pkg::*;
  `include "uvm_macros.svh"
  class rkv_sequence extends uvm_sequence;
    int var1, var2, var3;
    `uvm_object_utils(rkv_sequence)
    function new(string name = "rkv_sequence");
      super.new(name);
    endfunction
    task body();
      if(!uvm_config_db#(int)::get(m_sequencer, "", "var1", var1))
        `uvm_error("CFGDB", "Cannot get var1")
      else
        `uvm_info("CFGDB", $sformatf("var1 is %0d", var1), UVM_LOW)
      if(!uvm_config_db#(int)::get(uvm_root::get(),
                                   "uvm_test_top.seq", "var2", var2))
        `uvm_error("CFGDB", "Cannot get var2")
      else
        `uvm_info("CFGDB", $sformatf("var2 is %0d", var2), UVM_LOW)
      if(!uvm_config_db#(int)::get(null, "", "var3", var3))
        `uvm_error("CFGDB", "Cannot get var3")
      else
        `uvm_info("CFGDB", $sformatf("var3 is %0d", var3), UVM_LOW)
    endtask
  endclass

  class rkv_sequencer extends uvm_sequencer;
```

```
    `uvm_component_utils(rkv_sequencer)
    function new(string name = "rkv_sequencer",
                 uvm_component parent = null);
      super.new(name, parent);
    endfunction
  endclass

  class sequence_get_from_config_db_test extends uvm_test;
    rkv_sequencer sqr;
    `uvm_component_utils(sequence_get_from_config_db_test)
    function new(string name = "sequence_get_from_config_db_test",
                 uvm_ component parent = null);
      super.new(name, parent);
    endfunction
    function void build_phase(uvm_phase phase);
      uvm_config_db#(int)::set(this, "sqr","var1", 1);
      uvm_config_db#(int)::set(this, "seq","var2", 2);
      uvm_config_db#(int)::set(null, "","var3", 3);
      sqr = rkv_sequencer::type_id::create("sqr", this);
    endfunction
    task run_phase(uvm_phase phase);
      rkv_sequence seq = rkv_sequence::type_id::create("seq", this);;
      phase.raise_objection(this);
      seq.start(sqr);
      phase.drop_objection(this);
    endtask
  endclass
endpackage

module tb;
  import uvm_pkg::*;
  `include "uvm_macros.svh"
  import rkv_pkg::*;
  initial run_test("sequence_get_from_config_db_test");
endmodule
```

仿真结果：

```
    UVM_INFO @ 0: reporter [RNTST] Running test
    sequence_get_from_config_db_ test...
    UVM_INFO  uvm_sequence_get_from_config_db.sv(14)  @  0:  uvm_test_top.
sqr@@seq [CFGDB] var1 is 1
    UVM_INFO  uvm_sequence_get_from_config_db.sv(18)  @  0:  uvm_test_top.
sqr@@seq [CFGDB] var2 is 2
    UVM_INFO  uvm_sequence_get_from_config_db.sv(22)  @  0:  uvm_test_top.
sqr@@seq [CFGDB] var3 is 3
```

阅读手记：

3.3.4 start()和`uvm_do_on()有何区别？

start()是 uvm_sequence 的方法，而`uvm_do()`、`uvm_do_on()等一系列的宏只能在 uvm_sequence 中使用。

关键词：
start，`uvm_do_on

避坑指南：
在 uvm_test 中，如果要将顶层 virtual sequence 挂载到 virtual sequencer 上，则应该使用 *uvm_sequence::start()* 方法，而不是使用`uvm_do_on()`的宏，原因就是在 uvm_test 中无法使用`uvm_do_on()`的宏。

阅读手记：

3.3.5 uvm_sequence_library 的作用是什么？

在一般数据流的 sequence 处理中，*uvm_sequence_library（seq_lib）*"出镜"次数比较少。可以将它看作单元序列（element sequence）的"集合"。在定义 seq_lib 类时，配合使用`uvm_sequence_library_utils(SEQLIB)`。在实际使用中，可以例化 seq_lib 对象，将其挂载到指定 sequencer 上。该 seq_lib 会按照可选模式随机选定一个 sequence 执行，并且结合执行模式和次数进行随机排列。

关键词：
uvm_sequence_library，`uvm_add_to_seq_lib，`uvm_sequence_library_utils

避坑指南：
seq_lib 类定义的思路与 SystemVerilog randsequence 是类似的，它们都期望能够将随机的方式提高一个层面。不过，这两者也有特定的使用场景，尤其是在面向处理器的指令处理时（独立指令作为 sequence item），能更自如地运用。

参考代码： uvm_sequence_library_usage.sv

```systemverilog
package rkv_pkg;
  import uvm_pkg::*;
  `include "uvm_macros.svh"
  class rkv_sequencer extends uvm_sequencer;
    `uvm_component_utils(rkv_sequencer)
    function new(string name = "rkv_sequencer",
                 uvm_component parent = null);
      super.new(name, parent);
    endfunction
  endclass
  class rkv_driver extends uvm_driver;
    `uvm_component_utils(rkv_driver)
    function new(string name = "rkv_driver", uvm_component parent = null);
      super.new(name, parent);
    endfunction
  endclass
  class rkv_sequence_lib extends uvm_sequence_library;
    `uvm_object_utils(rkv_sequence_lib)
    `uvm_sequence_library_utils(rkv_sequence_lib)
    function new(string name = "rkv_sequence_lib");
      super.new(name);
      init_sequence_library();
    endfunction
  endclass
  class rkv_sequence1 extends uvm_sequence;
    `uvm_object_utils(rkv_sequence1)
    `uvm_add_to_seq_lib(rkv_sequence1, rkv_sequence_lib)
    function new(string name = "rkv_sequence1");
      super.new(name);
    endfunction
    task body();
      `uvm_info("SEQ",
        $sformatf("%s started now", get_type_name()), UVM_LOW)
    endtask
  endclass
  class rkv_sequence2 extends rkv_sequence1;
    `uvm_object_utils(rkv_sequence2)
    `uvm_add_to_seq_lib(rkv_sequence2, rkv_sequence_lib)
    function new(string name = "rkv_sequence2");
      super.new(name);
    endfunction
  endclass
  class rkv_sequence3 extends rkv_sequence1;
    `uvm_object_utils(rkv_sequence3)
    `uvm_add_to_seq_lib(rkv_sequence3, rkv_sequence_lib)
    function new(string name = "rkv_sequence3");
      super.new(name);
```

```
    endfunction
  endclass

  class sequence_library_usage_test extends uvm_test;
    rkv_driver driver;
    rkv_sequencer sequencer;
    uvm_sequence_library_cfg lib_cfg;
    `uvm_component_utils(sequence_library_usage_test)
    function new(string name = "sequence_library_usage_test",
                 uvm_ component parent = null);
      super.new(name, parent);
    endfunction
    function void build_phase(uvm_phase phase);
      driver = rkv_driver::type_id::create("driver", this);
      sequencer = rkv_sequencer::type_id::create("sequencer", this);
      lib_cfg = new("lib_cfg", UVM_SEQ_LIB_RANDC, 2, 5);
      uvm_config_db #(uvm_object_wrapper)::set(
                    this, "sequencer.run_ phase",
                    "default_sequence",
                    rkv_sequence_lib::get_type());
      uvm_config_db #(uvm_sequence_library_cfg)::set(
                    this, "sequencer. run_phase",
                    "default_sequence.config",
                    lib_cfg);
    endfunction
    function void connect_phase(uvm_phase phase);
      driver.seq_item_port.connect(sequencer.seq_item_export);
    endfunction
  endclass
endpackage

module tb;
  import uvm_pkg::*;
   `include "uvm_macros.svh"
  import rkv_pkg::*;
  initial run_test("sequence_library_usage_test");
endmodule
```

仿真结果：

```
    UVM_INFO @ 0: reporter [RNTST] Running test sequence_library_usage_test...
    UVM_INFO /$UVM_HOME/seq/uvm_sequence_library.svh(660) @ 0: uvm_test_top.
sequencer@@rkv_sequence_lib [SEQLIB/START] Starting sequence library rkv_sequence_
lib in run phase: 4 iterations in mode UVM_SEQ_LIB_RANDC
    UVM_INFO uvm_sequence_library_usage.sv(31) @ 0: uvm_test_top.sequencer
@@rkv_sequence_lib.rkv_sequence3:1 [SEQ] rkv_sequence3 started now
    UVM_INFO uvm_sequence_library_usage.sv(31) @ 0: uvm_test_top.sequencer
@@rkv_sequence_lib.rkv_sequence1:2 [SEQ] rkv_sequence1 started now
    UVM_INFO uvm_sequence_library_usage.sv(31) @ 0: uvm_test_top.sequencer@@
```

```
rkv_sequence_lib.rkv_sequence2:3 [SEQ] rkv_sequence2 started now
        UVM_INFO uvm_sequence_library_usage.sv(31) @ 0: uvm_test_top.sequencer@@
rkv_sequence_lib.rkv_sequence3:4 [SEQ] rkv_sequence3 started now
```

阅读手记：

3.3.6 配置 default_sequence 和调用 sequence::start()是否可同时进行？

请避免这种情况。这两种方式，都可以对目标 sequencer 配置其需要执行的 sequence。不过，default_sequence 即便做多次配置，最终也只允许一个生效，但 *sequence::start(SQR)* 不受此限制。

关键词：
default_sequence，sequence，start

避坑指南：
之所以要避免以上两种形式同时使用，是因为这两种方式指定的 sequence 都会被 sequencer 采纳，并且按照 sequencer 的仲裁模式进行发送。

阅读手记：

3.3.7 一些 sequence 调用 raise_objection()的目的是什么？

调用 raise_objection()的目的是防止当前所在的 phase 提前退出。一些 sequence 会为了防止执行时的 run_phase 或 main_phase 提前退出，可能在 *pre_body()*、*post_body()* 处分别调用 *raise_objection()*、*drop_objection()*。

关键词：
raise_objection，drop_objection

避坑指南：
如果每一个 sequence 带有 *raise_objection()*、*drop_objection()*，那么层层嵌套的防退出结构看起来是给保险柜上了多个"锁"，但也有可能变成解不开的"线团"。推荐在顶层 virtual sequence 执行时，使用 *sequence.start(SQR)* 这样的方式，只需一道"锁"就能够确保 sequence 的正常执行。注意，这时 *raise_objection()*、*drop_objection()* 在 test 层的 run_phase 或其他需要的 phase 中指定，而非在 sequence 中指定。

参考代码：uvm_sequence_top_objection.sv

```
package rkv_pkg;
  import uvm_pkg::*;
  `include "uvm_macros.svh"
  class rkv_sequence extends uvm_sequence;
    `uvm_object_utils(rkv_sequence)
    function new(string name = "rkv_sequence");
      super.new(name);
    endfunction
    task body();
      // available but not necessary to raise/drop objection inside sequence
      //if(phase != null) phase.raise_objection(this);
      `uvm_info("PHASE", "sequence body entered", UVM_LOW)
      #10ns;
      `uvm_info("PHASE", "sequence body exited", UVM_LOW)
      //if(phase != null) phase.drop_objection(this);
    endtask
  endclass

  class rkv_sequencer extends uvm_sequencer;
    `uvm_component_utils(rkv_sequencer)
    function new(string name = "rkv_sequencer",
                 uvm_component parent = null);
      super.new(name, parent);
    endfunction
  endclass

  class sequence_top_objection_test extends uvm_test;
    rkv_sequence seq1;
    rkv_sequencer sqr1;
    `uvm_component_utils(sequence_top_objection_test)
    function new(string name = "sequence_top_objection_test",
                 uvm_component parent = null);
      super.new(name, parent);
    endfunction
    function void build_phase(uvm_phase phase);
      super.build_phase(phase);
      seq1 = rkv_sequence::type_id::create("seq1", this);
      sqr1 = rkv_sequencer::type_id::create("sqr1", this);
    endfunction

    task run_phase(uvm_phase phase);
      super.run_phase(phase);
      `uvm_info("PHASE", "run phase entered", UVM_LOW)
      // stop virtial sequence raise/drop objection is a simple way
```

```
        phase.raise_objection(this);
        seq1.start(sqr1);
        phase.drop_objection(this);
        `uvm_info("PHASE", "run phase exited", UVM_LOW)
      endtask
    endclass
  endpackage

  module tb;
    import uvm_pkg::*;
    `include "uvm_macros.svh"
    import rkv_pkg::*;
    initial  run_test("sequence_top_objection_test");
  endmodule
```

仿真结果：

```
    UVM_INFO @ 0: reporter [RNTST] Running test sequence_top_objection_test...
    UVM_INFO uvm_sequence_top_objection.sv(42) @ 0: uvm_test_top [PHASE] run
phase entered
    UVM_INFO uvm_sequence_top_objection.sv(13) @ 0: uvm_test_top.sqr1@@seq1
[PHASE] sequence body entered
    UVM_INFO uvm_sequence_top_objection.sv(15) @ 10000: uvm_test_top.sqr1@@
seq1 [PHASE] sequence body exited
    UVM_INFO uvm_sequence_top_objection.sv(50) @ 10000: uvm_test_top [PHASE]
run phase exited
```

阅读手记：

3.3.8　每一个 sequence 都需要调用 raise_objection()吗？

不建议在每一个 sequence 中添加 *raise_objection()*、*drop_objection()*，尤其不建议在 *pre_body()*、*post_body()* 中添加，因为它们（*pre_body()*、*post_body()*）在 `uvm_do_xxx()` 有关的宏中并不会被调用。

同时，如果所有的 sequence 都添加了 objection 操作，则会使一些特殊情况中的复杂 sequence 调试变得异常困难（试想一下你给一道大门加装了很多锁）。所以，建议在顶层的 base virtual sequence 中添加 *raise_objection()*、*drop_objection()*，或在 base test 中添加 *raise_objection()*、*drop_objection()*。一个原则就是：只用一把锁来防止仿真提前退出。

关键词：

sequence，objection，raise，drop，test

避坑指南：

也许除了在 sequence、test 这些与测试场景有关的地方你想添加 *raise_objection()*、*drop_objection()*，在一些特定情况下，你还想在其他验证环境组件中添加 objection 操作。但我希望在能百分之百清楚你要做的事情和可能的收益之前，先考虑是否有其他办法利用 sequence、test 中的 objection 操作就能够达到目的，尽量让事情变得简单起来！多数情况下，一把锁就够了。

阅读手记：

3.3.9 set_automatic_phase_objection()使用起来方便吗？

多数情况下，*uvm_sequence::set_automatic_phase_objection()* 与 *raise_objection()*、*drop_objection()* 的成对操作的效果是一致的。*set_automatic_phase_objection()* 会在 sequence 进入 *start()* 后执行 raise objection 操作，在退出 *start()* 前执行 drop objection 操作。我们可以在 test 的 run_phase 挂起 sequence 前进行设置：

SEQ.set_starting_phase(phase);
SEQ.set_automatic_phase_objection(1);
SEQ.start(SQR);

通过这个设置，可以让 sequence 自动行使 objection 完成 raise 和 drop，事情看起来也变得简单起来。只不过这个新增的方法和原有的 objection 手动处理之间没有明显的优劣之别。*set_automatic_phase_objection()* 只在一些特定的情况下有影响，例如，要终止一个正在执行的 sequence，事先调用 *set_automatic_phase_objection()* 会更安全，防止可能出现的死锁问题。

关键词：

uvm_sequence，set_starting_phase，set_automatic_phase_objection，raise，drop

避坑指南：

遵循"一把锁"原则，对 *set_automatic_phase_objection()* 方法的调用，也只需要在 test 层添加即可。

参考代码： uvm_sequence_automatic_objection.sv

```
package rkv_pkg;
  import uvm_pkg::*;
  `include "uvm_macros.svh"
  class rkv_sequence extends uvm_sequence;
    `uvm_object_utils(rkv_sequence)
    function new(string name = "rkv_sequence");
      super.new(name);
```

```systemverilog
    endfunction
    task body();
      `uvm_info("PHASE", "sequence body entered", UVM_LOW)
      #10ns;
      `uvm_info("PHASE", "sequence body exited", UVM_LOW)
    endtask
  endclass

  class rkv_sequencer extends uvm_sequencer;
    `uvm_component_utils(rkv_sequencer)
    function new(string name = "rkv_sequencer",
                 uvm_component parent = null);
      super.new(name, parent);
    endfunction
  endclass

  class sequence_top_objection_test extends uvm_test;
    rkv_sequence seq1;
    rkv_sequencer sqr1;
    `uvm_component_utils(sequence_top_objection_test)
    function new(string name = "sequence_top_objection_test",
                 uvm_component parent = null);
      super.new(name, parent);
    endfunction
    function void build_phase(uvm_phase phase);
      super.build_phase(phase);
      seq1 = rkv_sequence::type_id::create("seq1", this);
      sqr1 = rkv_sequencer::type_id::create("sqr1", this);
    endfunction

    task run_phase(uvm_phase phase);
      super.run_phase(phase);
      `uvm_info("PHASE", "run phase entered", UVM_LOW)
      // sequence automatic phase objection also works
      seq1.set_starting_phase(phase);
      seq1.set_automatic_phase_objection(1);
      seq1.start(sqr1);
      `uvm_info("PHASE", "run phase exited", UVM_LOW)
    endtask
  endclass
endpackage

module tb;
  import uvm_pkg::*;
  `include "uvm_macros.svh"
  import rkv_pkg::*;
  initial run_test("sequence_top_objection_test");
endmodule
```

仿真结果：

```
UVM_INFO @ 0: reporter [RNTST] Running test sequence_top_objection_test...
UVM_INFO  uvm_sequence_automatic_objection.sv(38)  @  0:  uvm_test_top
[PHASE] run phase entered
UVM_INFO uvm_sequence_automatic_objection.sv(10) @ 0: uvm_test_top.sqr1
@@seq1 [PHASE] sequence body entered
UVM_INFO uvm_sequence_automatic_objection.sv(12) @ 10000: uvm_test_top.
sqr1@@seq1 [PHASE] sequence body exited
UVM_INFO uvm_sequence_automatic_objection.sv(43) @ 10000: uvm_test_top
[PHASE] run phase exited
```

阅读手记：

3.3.10 如何终止一个正在执行的 sequence？

你可能会考虑选择用 *disable THREAD* 这样的 SystemVerilog 原生的终止线程的方法，但事情并没有那么简单，因为有两个隐患没有排除。

第一，这个 sequence 线程可能要执行 raise objection，但你如果直接终止（disable）这个 sequence 线程，它将没有机会 drop objection；第二，这个 sequence 可能还在 sequencer 那里排队，并且在发送数据的过程中，如果直接 disable 它，它将无法归还 sequencer 的仲裁授权，使得 sequencer 此后无法正常与其他 sequence 保持 transaction 传输。

所以，安全的做法应该是，先利用 *uvm_sequence::set_automatic_phase_objection()* 在挂载 sequencer 之前设置自动的 objection 机制，接下来在需要终止 sequence 的地方调用 *uvm_sequence::kill()* 方法。该方法会配合 automatic objection 机制，确保及时完成 drop objection 操作，同时还将其从所挂载的 sequencer 中移出，避免死锁情况。

关键词：

uvm_sequence，disable，objection，set_automatic_phase_objection，kill

避坑指南：

如果 sequence 线程本身没有使用 objection 机制，那么也就不需要调用 *SEQ.set_automatic_phase_objection(1)*。

参考代码： uvm_sequence_kill.sv

```
package rkv_pkg;
  import uvm_pkg::*;
  `include "uvm_macros.svh"
  class rkv_sequence extends uvm_sequence;
```

```systemverilog
`uvm_object_utils(rkv_sequence)
function new(string name = "rkv_sequence");
  super.new(name);
endfunction
task body();
  `uvm_info("PHASE", "sequence body entered", UVM_LOW)
  #10ns;
  `uvm_info("PHASE", "sequence body exited", UVM_LOW)
endtask
endclass

class rkv_sequencer extends uvm_sequencer;
  `uvm_component_utils(rkv_sequencer)
  function new(string name = "rkv_sequencer",
               uvm_component parent = null);
    super.new(name, parent);
  endfunction
endclass

class sequence_kill_test extends uvm_test;
  rkv_sequence seq1, seq2;
  rkv_sequencer sqr;
  `uvm_component_utils(sequence_kill_test)
  function new(string name = "sequence_kill_test",
               uvm_component parent = null);
    super.new(name, parent);
  endfunction
  function void build_phase(uvm_phase phase);
    super.build_phase(phase);
    seq1 = rkv_sequence::type_id::create("seq1", this);
    seq2 = rkv_sequence::type_id::create("seq2", this);
    sqr  = rkv_sequencer::type_id::create("sqr", this);
  endfunction

  task run_phase(uvm_phase phase);
    super.run_phase(phase);
    phase.raise_objection(this);
    `uvm_info("PHASE", "run phase entered", UVM_LOW)
    fork
      #5ns seq1.kill();
    join_none
    seq1.start(sqr);
    seq2.start(sqr);
    `uvm_info("PHASE", "run phase exited", UVM_LOW)
    phase.drop_objection(this);
  endtask
endclass
```

```
    endpackage

module tb;
    import uvm_pkg::*;
    `include "uvm_macros.svh"
    import rkv_pkg::*;
    initial run_test("sequence_kill_test");
endmodule
```

仿真结果:

```
    UVM_INFO @ 0: reporter [RNTST] Running test sequence_kill_test...
    UVM_INFO uvm_sequence_kill.sv(40) @ 0: uvm_test_top [PHASE] run phase entered
    UVM_INFO uvm_sequence_kill.sv(10) @ 0: uvm_test_top.sqr@@seq1 [PHASE] sequence body entered
    UVM_INFO uvm_sequence_kill.sv(10) @ 5000: uvm_test_top.sqr@@seq2 [PHASE] sequence body entered
    UVM_INFO uvm_sequence_kill.sv(12) @ 15000: uvm_test_top.sqr@@seq2 [PHASE] sequence body exited
    UVM_INFO uvm_sequence_kill.sv(46) @ 15000: uvm_test_top [PHASE] run phase exited
```

阅读手记:

3.3.11 发送 sequence 和 sequence item 的优先级问题是什么?

uvm_sequence_base(sequence 的父类)具备成员变量 m_priority(优先级),该属性默认继承 sequence 所在的父一级的 sequence,也默认传送至它所例化、挂载的子一级 sequence。在传送 sequence 或 sequence item 时,可以分别通过方法 *uvm_sequence::start()* 或 *uvm_sequence_item::start_item()* 的参数将指定的优先级传入,继而影响该 sequence、sequence item 接下来的 sequencer 一侧的仲裁表现。如果在调用以上方法时使用了默认的优先级-1,那么所挂载的子一级 sequence、sequence item 将沿用父一级的优先级。

关键词:

uvm_sequence,uvm_sequence_item,priority,start,start_item

避坑指南:

使用诸如 `uvm_do_pri()`、`uvm_do_pri_with()`、`uvm_do_on_pri()`、`uvm_do_on_pri_with()` 这样的宏,也可以在发送子一级 sequence、sequence item 时指定优先级。

```
// uvm_sequence_base.svh
```

```
virtual task start (uvm_sequencer_base sequencer,
                   uvm_sequence_base parent_sequence = null,
                   int this_priority = -1,
                   bit call_pre_post = 1);
   ...
endtask
virtual task start_item (uvm_sequence_item item,
                        int set_priority = -1,
                        uvm_sequencer_base sequencer=null);
   ...
endtask
```

阅读手记：

3.3.12 为什么 sequence 通过 get_response()可以得到正确的 response？

无论是通过 `uvm_do 等宏的方式，还是通过 *start_item()*、*finish_item()*的方式来发送 sequence item 到 driver 一侧，最终 driver 可以将总线的数据和状态通过 *put_response()*或 *item_done()*方法返回至 sequence。可能有多个 sequence 在同一时间挂载到 sequencer，driver 会先后从多个 sequence 处获得不同的 sequence item。driver 在返回 response item 前，可以通过 *uvm_sequene_item::{set_sequence_id(), set_transaction_id()}*将 response item 与 request item 做逐一关联。在 response item 具备了与 request item 相同的 sequence id 和 transaction id 之后，它们最终将通过 sequencer 到达目标 sequence 处，继而 sequence 可以获得正确的 response item。

关键词：

put_response，item_done，set_sequence_id，set_transaction_id，get_response

避坑指南：

对于复杂总线的 driver，不一定会按照获得 request item 的顺序来进行驱动，也未必会按照 request item 的顺序来返回 response item。针对这种情况，uvm_sequence 一侧在调用 *get_response(output RSP response, input int transaction_id = -1)*时，需要考虑指定 transaction_id 这个参数，以便能够获得与目标 request item 的 id 相匹配的 response item。

阅读手记：

3.3.13 通过 uvm_config_db::set()或 start()挂载 sequence 有哪些联系和差别？

这两种方式都可以挂载 sequence，但需要注意这两种方式的不同用法。更倾向于建议采用 *uvm_sequence::start()* 方法来挂载 sequence，因为这种方式操作起来更为直观，而且 objection 的 raise、drop 操作也只需要在 *uvm_test::run_phase()* 中调用，并且在挂载前还可以做出赋值、随机化等操作。

也可以通过 *uvm_config_db::set()* 方式挂载 default_sequence。如果在挂载前要控制 sequence 实例中的数据（赋值、随机化），为了避免 sequencer 再次对其进行随机化，需要在配置前将 *uvm_sequence::do_not_randomize* 设置为 1。同时，需要考虑预先在 sequence 中做 objection 的 raise、drop 操作。这里推荐在 *uvm_sequence::new()* 函数中，通过 *uvm_sequence::set_automatic_phase_objection(1)* 进行自动的 objection 操作（即不再需要手动完成 objection 的 raise、drop 操作）。

关键词：
uvm_sequence，uvm_config_db，start，objection，randomize，set_automatic_phase_objection

避坑指南：
避免混合使用这两种方法来挂载 sequence，否则将增加调试的难度。尽可能在整个项目中贯穿使用一种方法，可以通过在不同 phase 中来挂载 sequence（无论采用这两种方法的哪一种），实现测试序列的有序安排以及可能的跳转。而在底层组件如 driver 中，则只需要使用 run_phase 来使其保持运转即可，且它仍然能够在不同 phase 中（例如，reset_phase、main_phase）获得 sequence item 进行驱动。也应该避免在挂载 sequence 时同时出现 run_phase 和 main_phase 的挂载，尽管这样做不违反规定，但会给调试和理解测试序列带来困惑。

参考代码：uvm_sequence_start_vs_config.sv

```
package rkv_pkg;
  import uvm_pkg::*;
  `include "uvm_macros.svh"
  class seqfoo extends uvm_sequence;
    rand int data = -1;
    `uvm_object_utils(seqfoo)
    function new(string name = "seqfoo");
      super.new(name);
      set_automatic_phase_objection(1);
    endfunction
    task body();
      `uvm_info(get_type_name(), $sformatf("data is %0d", data), UVM_LOW)
      repeat(2) #1ns `uvm_do(req)
    endtask
  endclass

  class sequencer extends uvm_sequencer;
```

```
  `uvm_component_utils(sequencer)
  function new(string name = "sequencer", uvm_component parent = null);
    super.new(name, parent);
  endfunction
endclass

class driver extends uvm_driver;
  `uvm_component_utils(driver)
  function new(string name = "driver", uvm_component parent = null);
    super.new(name, parent);
  endfunction
  task run_phase(uvm_phase phase);
    get_and_drive("run_phase");
  endtask
  task get_and_drive(string ph);
    forever begin
      seq_item_port.get_next_item(req);
      `uvm_info(get_type_name(),
        $sformatf("At [%s] got next item from sequence [%s]", ph,
          req.get_parent_sequence().get_name()), UVM_LOW)
      seq_item_port.item_done();
    end
  endtask : get_and_drive
endclass

class sequence_start_vs_config_test extends uvm_test;
  driver drv;
  sequencer sqr;
  `uvm_component_utils(sequence_start_vs_config_test)

  function new(string name = "sequence_start_vs_config_test",
               uvm_component parent = null);
    super.new(name, parent);
  endfunction
  function void build_phase(uvm_phase phase);
    seqfoo s0 = seqfoo::type_id::create("seq0");
    seqfoo s1 = seqfoo::type_id::create("seq1");
    sqr = sequencer::type_id::create("sqr", this);
    drv = driver::type_id::create("drv", this);
    s0.randomize() with {data == 10;};
    // prevents randomization as a default sequence
    s0.do_not_randomize = 1;
    s1.randomize() with {data == 20;};
    // prevents randomization as a default sequence
    s1.do_not_randomize = 1;
    uvm_config_db #(uvm_sequence_base)::set(
                 this, "sqr.run_phase","default_sequence",s0);
```

```
            uvm_config_db #(uvm_sequence_base)::set(
                       this, "sqr.main_phase","default_sequence",s1);
        endfunction
        function void connect_phase(uvm_phase phase);
          drv.seq_item_port.connect(sqr.seq_item_export);
        endfunction
        task run_phase(uvm_phase phase);
          seqfoo s2 = seqfoo::type_id::create("seq2");
          phase.raise_objection(this);
          s2.randomize() with {data == 30;};
          s2.start(sqr);
          phase.drop_objection(this);
        endtask
      endclass
    endpackage

    module tb;
      import uvm_pkg::*;
      `include "uvm_macros.svh"
      import rkv_pkg::*;
      initial run_test("sequence_start_vs_config_test");
    endmodule
```

仿真结果：

```
        UVM_INFO @ 0: reporter [RNTST] Running test
        sequence_start_vs_config_ test...
        UVM_INFO uvm_sequence_start_vs_config.sv(12) @ 0: uvm_test_top.sqr@@seq0
[seqfoo] data is 10
        UVM_INFO uvm_sequence_start_vs_config.sv(12) @ 0: uvm_test_top.sqr@@seq2
[seqfoo] data is 30
        UVM_INFO uvm_sequence_start_vs_config.sv(12) @ 0: uvm_test_top.sqr@@seq1
[seqfoo] data is 20
        UVM_INFO uvm_sequence_start_vs_config.sv(35) @ 1000: uvm_test_top.drv
[driver] At [run_phase] got next item from sequence [seq0]
        UVM_INFO uvm_sequence_start_vs_config.sv(35) @ 1000: uvm_test_top.drv
[driver] At [run_phase] got next item from sequence [seq2]
        UVM_INFO uvm_sequence_start_vs_config.sv(35) @ 1000: uvm_test_top.drv
[driver] At [run_phase] got next item from sequence [seq1]
        UVM_INFO uvm_sequence_start_vs_config.sv(35) @ 2000: uvm_test_top.drv
[driver] At [run_phase] got next item from sequence [seq0]
        UVM_INFO uvm_sequence_start_vs_config.sv(35) @ 2000: uvm_test_top.drv
[driver] At [run_phase] got next item from sequence [seq2]
        UVM_INFO uvm_sequence_start_vs_config.sv(35) @ 2000: uvm_test_top.drv
[driver] At [run_phase] got next item from sequence [seq1]
```

阅读手记：

3.3.14 通过 uvm_config_db 挂载 default sequence 需要注意什么？

可以通过 *uvm_config_db #(uvm_sequence_base)::set()* 的方式挂载某一个 sequence 实例，在挂载前可以对该实例进行赋值或随机化。也可以通过 *uvm_config_db #(uvm_object_wrapper)::set()* 来挂载某一个注册在工厂中的类，挂载成功后在 UVM 代码内部做例化和随机化处理。第二种挂载形式便于后期的类型覆盖，实现灵活修改序列内容的效果。

关键词：

default_sequence，uvm_config_db，uvm_sequence_base，uvm_object_wrapper

避坑指南：

default_sequence 的设置最终会按照 *uvm_config_db::set()* 的配置优先级逻辑选取一个 sequence。所以，在配置 default_sequence 时，注意高层次对低层次的优先级、后配置对前配置的优先级。

参考代码： uvm_sequence_config_diff_ways.sv

```
package rkv_pkg;
  import uvm_pkg::*;
  `include "uvm_macros.svh"
  class seqfoo extends uvm_sequence;
    rand int data = -1;
    `uvm_object_utils(seqfoo)
    function new(string name = "seqfoo");
      super.new(name);
      set_automatic_phase_objection(1);
    endfunction
    task body();
      `uvm_info(get_type_name(), $sformatf("data is %0d", data), UVM_LOW)
      repeat(2) #1ns `uvm_do(req)
    endtask
  endclass

  class sequencer extends uvm_sequencer;
    `uvm_component_utils(sequencer)
    function new(string name = "sequencer", uvm_component parent = null);
      super.new(name, parent);
    endfunction
  endclass
```

```systemverilog
class driver extends uvm_driver;
  `uvm_component_utils(driver)
  function new(string name = "driver", uvm_component parent = null);
    super.new(name, parent);
  endfunction
  task run_phase(uvm_phase phase);
    get_and_drive("run_phase");
  endtask
  task get_and_drive(string ph);
    forever begin
      seq_item_port.get_next_item(req);
      `uvm_info(get_type_name(),
                $sformatf("At [%s] got next item from sequence [%s]",
                          ph, req.get_parent_sequence().get_name()),
                UVM_LOW)
      seq_item_port.item_done();
    end
  endtask : get_and_drive
endclass

class sequence_config_diff_ways_test extends uvm_test;
  driver drv;
  sequencer sqr;
  `uvm_component_utils(sequence_config_diff_ways_test)
  function new(string name = "sequence_config_diff_ways_test",
               uvm_component parent = null);
    super.new(name, parent);
  endfunction
  function void build_phase(uvm_phase phase);
    seqfoo s0 = seqfoo::type_id::create("seq0");
    seqfoo s1 = seqfoo::type_id::create("seq1");
    sqr = sequencer::type_id::create("sqr", this);
    drv = driver::type_id::create("drv", this);
    s0.randomize() with {data == 10;};
    // prevents randomization as a default sequence
    s0.do_not_randomize = 1;
    s1.randomize() with {data == 20;};
    // prevents randomization as a default sequence
    s1.do_not_randomize = 1;
    uvm_config_db #(uvm_sequence_base)::set(
                  this, "sqr.run_phase","default_sequence",s0);
    uvm_config_db #(uvm_sequence_base)::set(
                  this, "sqr.run_phase","default_sequence",s1);
    uvm_config_db #(uvm_object_wrapper)::set(
                  this, "sqr.run_phase","default_sequence",
                  seqfoo::type_id::get());
```

```
      endfunction
      function void connect_phase(uvm_phase phase);
        drv.seq_item_port.connect(sqr.seq_item_export);
      endfunction
    endclass
endpackage

module tb;
  import uvm_pkg::*;
  `include "uvm_macros.svh"
  import rkv_pkg::*;
  initial run_test("sequence_config_diff_ways_test");
endmodule
```

仿真结果：

```
UVM_INFO @ 0: reporter [RNTST] Running test
sequence_config_diff_ways_test...
UVM_INFO uvm_sequence_config_diff_ways.sv(12) @ 0: uvm_test_top.sqr@@
seqfoo [seqfoo] data is -661565555
UVM_INFO uvm_sequence_config_diff_ways.sv(35) @ 1000: uvm_test_top.drv
[driver] At [run_phase] got next item from sequence [seqfoo]
UVM_INFO uvm_sequence_config_diff_ways.sv(35) @ 2000: uvm_test_top.drv
[driver] At [run_phase] got next item from sequence [seqfoo]
```

阅读手记：

3.3.15 为什么不建议使用 default_sequence 挂载顶层序列呢？

用户在挂载 sequence 时希望在 run_phase、main_phase 这样的地方来完成，但 default_sequence 的设置方式是在 build_phase 中，这往往与我们对测试执行的理解有出入。

此外，通过 default_sequence 设置的方式，不容易控制什么时间开始执行 default_sequence，也不能很方便地挂载多个 sequence。对于不熟悉 default_sequence 的创建、随机化、配置等流程的用户来说，稍不谨慎的使用又会带来调试方面的麻烦。因为 default_sequence 从配置到执行的过程，其实在 UVM 源代码中隐藏了很多细节。

default_sequence 在 OVM 时代是较为通用和被建议的方式，但这一方式到了 UVM 时代已经被 uvm_sequence::start() 这一更直观、更容易控制的挂载方式取代了。尽管 default_sequence 并没有在 UVM 源代码中全部被摒弃掉（deprecated），但希望用户在构建新的测试用例时，尽可能使用风格统一的 uvm_sequence::start() 方式。

关键词：
default_sequence，phase，uvm_config_db，start

避坑指南：
用户在遇到采用 default_sequence 挂载顶层序列的时候，容易产生错觉，那就是这种方式动不得、只能保留、不能修改，但其实它可以在 test 层的 run_phase、main_phase 任务内简单地转换为 *uvm_sequence::start()* 方式。这种转换的成本并不高，而转换的好处是更加便于对顶层 sequence 的控制、随机化和挂载调试。

阅读手记：

3.3.16 uvm_sequence::start()挂载的 sequencer 什么情况下需要指定？

uvm_sequence 一旦挂载到 sequencer，那么这个 sequence 的 m_sequencer 句柄（以及 p_sequencer 句柄）都将指向到目标 sequencer。*uvm_sequence::start()* 挂载 sequencer 的作用就是能够找到一个挂载的目标。例如，virtual sequence 通过 start()指向到 virtual sequencer，可以通过 p_sequencer 路由找到 virtual sequencer 内部的其他成员；又如，element sequence 通过 start()指向目标 sequencer，就可以将其内部创建的 sequence item 送往 sequencer。那么，uvm_sequence::start()什么情况下不需要指向 sequencer 呢？就是 *uvm_sequence::start(null)*。一般有两种情况。第一种情况，这个 sequence 不需要目标 sequencer 句柄，也不会通过它访问 sequencer 的各个成员；第二种情况，这个 sequence 虽然没有指定挂载的 sequencer 句柄，但是可以通过其 parent sequence 挂载的 sequencer 句柄发送 sequence item。

关键词：
uvm_sequence，start，uvm_sequencer，null

避坑指南：
在通过 RGM 模型进行寄存器访问时（如 *uvm_reg::{write, read}*），其所在的 sequence 可以不指定 sequencer。原因是寄存器访问的操作是通过在 RGM、adapter、sequencer 之间完成的桥接转换，并不需要 *uvm_sequence::start()* 去指定 sequencer 来完成。

阅读手记：

3.3.17 virtual sequence 需要获得某些信号和状态应该如何实现？

virtual sequence 为了调度各个 sequence，除了需要各个 sequence 内部的成员、状态来协助，可能还需要测试平台中的某些信号和状态。这些信号状态或来自于设计，或来自于测试平台的其他组件。

如果是监测来自设计的某些信号状态，建议先将信号从设计内部"拉出"到接口中，由接口统一收集这些目标信号，再将接口传递到 sequence。这样，virtual sequence 最终可以获得接口句柄和内部的信号。

如果是要获得总线上面的某些事务状态，例如数据传输的开始和结束，那么可以利用验证组件中的 uvm_event。例如，在 monitor 内部创建和使用 uvm_event，那么其他组件和 sequence 可以先获得这些事件的句柄，再对它们进行监听，并接收 uvm_event 触发的数据内容。

关键词：
virtual sequence，virtual sequencer，virtual interface，uvm_event

避坑指南：
从组件和测试内容的独立性要求来看，应该尽量减少 sequence 与设计层次、验证层次的耦合。比如，避免通过 sequence 直接访问跨设计层次的变量，或者避免访问跨验证层次的某个组件的变量。这些处理方式在日后都将对测试用例的复用带来帮助。

阅读手记：

3.3.18 怎么让 sequence 感知 coverage 的增长并及时停止呢？

SystemVerilog 并没有将随机产生数值与覆盖率映射做关联的规则，这也为目前一些 EDA 工具的开发带来了机会和发展空间。为了让 sequence 能够控制随机值的生成和激励数量，我们可以先在 coverage model 上做一些必要的工作，包括控制 coverage model 的采样，控制各个 covergroup 的采样，获得个别 covergroup 的覆盖率和整体的覆盖率等必要接口。在测试序列产生随机数值以后，monitor 会将数据通过 analysis port 送往 coverage model，而 coverage model 会对其进行采样。这样，sequence 可以定期通过 coverage model 获得目标覆盖率状态，在覆盖率收集完成或到达最大激励数据（或最长仿真时间）之后，可以停止发送激励，结束测试。

关键词：
sequence，coverage，covergroup，randomization

避坑指南：

如果缺少必要的工具或方法让随机数据的产生更为有效，那么这一方式只能解决控制激励数量的问题。即便如此，把这一原则在项目间贯彻下来，也依然能够减少很多不必要的激励数据。

参考代码： uvm_sequence_control_with_coverage.sv

```
package rkv_pkg;
  import uvm_pkg::*;
  `include "uvm_macros.svh"
  class user_config extends uvm_object;
    bit coverage_enable = 1;
    bit covergroup_enable [string];
    `uvm_object_utils(user_config)
    function new(string name = "user_config");
      super.new(name);
    endfunction
  endclass

  class user_item extends uvm_object;
    randc bit [1:0] d1, d2;
    `uvm_object_utils(user_item)
    function new(string name = "user_item");
      super.new(name);
    endfunction
  endclass

  covergroup user_cg with function sample(bit[1:0] d);
    option.per_instance = 1;
    coverpoint d;
  endgroup

  class cov_model extends uvm_subscriber #(user_item);
    user_config cfg;
    user_cg cg1;
    user_cg cg2;
    `uvm_component_utils(cov_model)

    function new(string name = "cov_model", uvm_component parent = null);
      super.new(name, parent);
      cg1 = new();
      cg2 = new();
    endfunction

    function void end_of_elaboration_phase(uvm_phase phase);
      cfg.covergroup_enable["cg1"] = 1;
      cfg.covergroup_enable["cg2"] = 1;
```

```
    endfunction

    function void write(T t);
      // sample data from argument
      foreach(cfg.covergroup_enable[idx]) begin
        if(!is_covergroup_enabled(idx))
          continue;
        case(idx)
          "cg1"  : cg1.sample(t.d1);
          "cg2"  : cg2.sample(t.d2);
        endcase
      end
    endfunction

    function bit is_covergroup_enabled(string name);
      case(name)
        "cg1"    : return cfg.coverage_enable &&
                          cfg.covergroup_enable ["cg1"];
        "cg2"    : return cfg.coverage_enable &&
                          cfg.covergroup_enable ["cg2"];
        default  : `uvm_error("UNDEF",
                     $sformatf("unrecognized coverage name %s",name))
      endcase
    endfunction

    function real get_coverage(string name);
      case(name)
        "cg1"    : get_coverage = cg1.get_inst_coverage();
        "cg2"    : get_coverage = cg2.get_inst_coverage();
        default  : `uvm_error("UNDEF",
                     $sformatf("unrecognized coverage name %s",name))
      endcase
       `uvm_info("COVRPT",
          $sformatf("%s coverage = %.1f", name, get_coverage), UVM_LOW)
    endfunction

    function bit is_coverage_complete(string names[]);
      foreach(names[i]) begin
        if(get_coverage(names[i]) < 100.0)
          return 0;
      end
      return 1;
    endfunction
endclass

class sequence_control_with_coverage_test extends uvm_test;
  user_config cfg;
```

```
      cov_model cm;
      `uvm_component_utils(sequence_control_with_coverage_test)
      function new(string name = "sequence_control_with_coverage_test",
                  uvm_component parent = null);
        super.new(name, parent);
      endfunction
      function void build_phase(uvm_phase phase);
        cfg = user_config::type_id::create("cfg", this);
        cm = cov_model::type_id::create("cm", this);
      endfunction
      function void connect_phase(uvm_phase phase);
        cm.cfg = cfg;
      endfunction
      task run_phase(uvm_phase phase);
        user_item t = user_item::type_id::create("t");
        phase.raise_objection(this);
        `uvm_info("COVRPT", "coverage initial report::", UVM_LOW)
        void'(cm.get_coverage("cg1"));
        void'(cm.get_coverage("cg2"));
        forever begin
          void'(t.randomize());
          `uvm_info("COVSMP", "coverage is sampled", UVM_LOW)
          cm.write(t);
          if(cm.is_coverage_complete('{"cg1", "cg2"}))
            break;
        end
        `uvm_info("COVRPT", "coverage finish report::", UVM_LOW)
        void'(cm.get_coverage("cg1"));
        void'(cm.get_coverage("cg2"));
        phase.drop_objection(this);
      endtask
    endclass
  endpackage

  module tb;
    import uvm_pkg::*;
    `include "uvm_macros.svh"
    import rkv_pkg::*;
    initial run_test("sequence_control_with_coverage_test");
  endmodule
```

仿真结果：

```
    UVM_INFO @ 0: reporter [RNTST] Running test coverage_control_test...
    UVM_INFO uvm_sequence_control_with_coverage.sv(98) @ 0: uvm_test_top
[COVRPT] coverage initial report::
    UVM_INFO uvm_sequence_control_with_coverage.sv(69) @ 0: uvm_test_top.cm
[COVRPT] cg1 coverage = 0.0
```

```
        UVM_INFO uvm_sequence_control_with_coverage.sv(69) @ 0: uvm_test_top.cm
[COVRPT] cg2 coverage = 0.0
        UVM_INFO uvm_sequence_control_with_coverage.sv(103) @ 0: uvm_test_top
[COVSMP] coverage is sampled
        UVM_INFO uvm_sequence_control_with_coverage.sv(69) @ 0: uvm_test_top.cm
[COVRPT] cg1 coverage = 25.0
        UVM_INFO uvm_sequence_control_with_coverage.sv(103) @ 0: uvm_test_top
[COVSMP] coverage is sampled
        UVM_INFO uvm_sequence_control_with_coverage.sv(69) @ 0: uvm_test_top.cm
[COVRPT] cg1 coverage = 50.0
        UVM_INFO uvm_sequence_control_with_coverage.sv(103) @ 0: uvm_test_top
[COVSMP] coverage is sampled
        UVM_INFO uvm_sequence_control_with_coverage.sv(69) @ 0: uvm_test_top.cm
[COVRPT] cg1 coverage = 75.0
        UVM_INFO uvm_sequence_control_with_coverage.sv(103) @ 0: uvm_test_top
[COVSMP] coverage is sampled
        UVM_INFO uvm_sequence_control_with_coverage.sv(69) @ 0: uvm_test_top.cm
[COVRPT] cg1 coverage = 100.0
        UVM_INFO uvm_sequence_control_with_coverage.sv(69) @ 0: uvm_test_top.cm
[COVRPT] cg2 coverage = 100.0
        UVM_INFO uvm_sequence_control_with_coverage.sv(108) @ 0: uvm_test_top
[COVRPT] coverage finish report::
        UVM_INFO uvm_sequence_control_with_coverage.sv(69) @ 0: uvm_test_top.cm
[COVRPT] cg1 coverage = 100.0
        UVM_INFO uvm_sequence_control_with_coverage.sv(69) @ 0: uvm_test_top.cm
[COVRPT] cg2 coverage = 100.0
```

阅读手记：

3.4 寄存器模型

在建立寄存器模型并将其与 adapter、sequencer、predictor 等集成后，就可以使用它进行寄存器有关的访问了。在构建和使用寄存器模型时也会有一些问题，例如，寄存器创建时的参数给定、访问时的请求和反馈调试，以及有关的回调函数使用等。对于层次化的寄存器模型，在遇到寄存器模型层次嵌套、地址信息映射管理、数据位宽差异等问题时，也会有诸多疑难点。接下来我们将就这些话题逐一展开讨论。

3.4.1 寄存器模型验证常见的测试点有哪些？

第一，检查寄存器的复位值，第二，检查寄存器域的常见读写属性（RO/RW/WO 等），第三，检查每个寄存器的地址映射关系是否正确。这三者均可通过 UVM 寄存器模型内建测试序列完成（对个别寄存器或域可将其排除对应的序列检查）。第四，检查状态寄存器的反馈是否及时准确，这一点需要检查硬件状态信号是否连接到寄存器端，如果更新方式是主动更新，那么可通过后门访问进行快速检查（不占用总线）；如果更新方式是被动更新，那么只能通过前门访问触发硬件状态值更新。这两种方式都需要将内部监测到的硬件信号与读取的值进行比较。第五，对于一些特殊寄存器（write-clear、read-clear、write-only-once），需要结合其特定属性进行单独访问，并且通过后门访问或监测内部信号，检查其功能。

关键词：
register model，field，built-in sequence，backdoor，frontdoor，monitor

避坑指南：
在寄存器验证过程中，后门路径设定是必需的一个环节，否则无法进行上述地址映射检查和特殊寄存器的检查，也无法做快速的后门读写访问。

阅读手记：

3.4.2 使用 set_auto_predict() 和 predictor 更新寄存器有什么区别？

如果是后门读写访问，那么会直接在 uvm_reg 访问结束时，调用 do_predict() 更新 uvm_reg 的镜像值、期望值。如果是前门访问，那么需要设置 *uvm_reg_map::set_auto_predict()*，或依靠 predictor，两者需要总线数据去更新 uvm_reg 的镜像值、期望值。

关键词：
uvm_reg，set_auto_predict，predictor，mirrored value，desired value

避坑指南：

uvm_reg_map 默认情况下没有启用 auto predict 模式。如果寄存器访问均依赖寄存器模型，那么可调用 *set_auto_predict()* 完成寄存器期望值、镜像值的更新；如果寄存器访问有多个主端（并非只通过寄存器模型），或直接通过总线模型，那么应该使用 predictor，以免 auto predict 模式遗漏其他无法监测到的寄存器访问。

阅读手记：

3.4.3 如何对寄存器的某些域进行读/写操作？

在利用寄存器模型对寄存器进行读/写访问时，往往对整个寄存器进行读/写的机会不多，更多的是修改某些单独的域，或获得某些域的值。这意味着利用 *uvm_reg::{write, read}* 的机会不多，更多的是利用 *uvm_reg_field::{set, get}*；后者也更易于阅读和维护。

例如，对于某寄存器的若干域进行写操作：

RGM.REGx.FIELD1.set(DATA1);
RGM.REGx.FIELD2.set(DATA2);
RGM.REGx.update(status);

又如，获取某寄存器的某个域的值：

RGM.REGx.mirror(status);
DATA = RGM.REGx.FIELD1.get();

关键词：

register model，field，write，read，set，get

避坑指南：

要利用好 *uvm_reg_field::{set,get}* 的操作，就一定要注意及时调用 *uvm_reg::{update, mirror}*，确保寄存器的值进行过读写，继而保证寄存器模型内的寄存器值与设计中的寄存器值保持一致。

阅读手记：

3.4.4 uvm_reg_filed::configure()中的参数 volatile 的作用是什么？

根据 UVM 用户手册中的说明，在构建寄存器模型过程中给个别域配置属性时，设置 *volatile=1*，表明在访问该域时它的值是无法预期的。比如，如果典型的域访问属性是 RO（Read Only）或 WO（Write Only），那么这些域的值在下一次访问时都是不可预测的（既不能通过前门预测，也不能通过后门访问时伴随的镜像值更新来预测）。

```
// uvm_ref_field.svh

// Function: is_volatile
// Indicates if the field value is volatile
// UVM uses the IEEE 1685-2009 IP-XACT definition of "volatility".
// If TRUE, the value of the register is not predictable because it
// may change between consecutive accesses.
// This typically indicates a field whose value is updated by the DUT.
// The nature or cause of the change is not specified.
// If FALSE, the value of the register is not modified between
// consecutive accesses.

extern virtual function bit is_volatile();
```

所以，volatile 影响寄存器行为的第一点，就是在调用 *uvm_reg::update()* 函数时，无论期望值与镜像值是否一致，都将触发一次对寄存器的访问。这对于诸如使能某种功能或清除中断等需要触发总线访问寄存器的操作是有效的。

```
// uvm_reg_field.svh

// Function: needs_update
// Check if abstract model contains different desired and mirrored values.
// If a desired field value has been modified in the abstraction class
// without actually updating the field in the DUT,
// the state of the DUT (more specifically what the abstraction class
// ~thinks~ the state of the DUT is) is outdated.
// This method returns TRUE
// if the state of the field in the DUT needs to be updated
// to match the desired value.
// The mirror values or actual content of DUT field are not modified.
// Use the <uvm_reg::update()> to actually update the DUT field.

function bit uvm_reg_field::needs_update();
   needs_update = (m_mirrored != m_desired) | m_volatile;
endfunction: needs_update
```

volatile 影响寄存器行为的第二点，就是在调用 *uvm_reg::mirror()* 函数时，默认将镜像值与读取的值进行比较。但是，如果该域的属性 m_check 是 UVM_NO_CHECK，那么将不会做比较，而该属性又是直接由 m_volatile 影响的。此时，如果某个域的属性为 RO（状态寄存器属性），那么仍然可以通过 mirror 方法获得该状态寄存器的值，而不会对该域的值做检查（即使 mirror 方法调用时的参数 check=UVM_CHECK，也不会检查该域的值）。

```
// uvm_reg_field.svh
function void uvm_reg_field::configure(uvm_reg        parent,
                                       int unsigned   size,
                                       int unsigned   lsb_pos,
                                       string         access,
                                       bit            volatile,
                                       uvm_reg_data_t reset,
                                       bit            has_reset,
                                       bit            is_rand,
                                       bit            individually_accessible);
    m_parent = parent;
    if (size == 0) begin
      `uvm_error("RegModel",
         $sformatf("Field \"%s\" cannot have 0 bits", get_full_name()));
      size = 1;
    end

    m_size      = size;
    m_volatile  = volatile;
    m_access    = access.toupper();
    m_lsb       = lsb_pos;
    m_cover_on  = UVM_NO_COVERAGE;
    m_written   = 0;
    m_check     = volatile ? UVM_NO_CHECK : UVM_CHECK;
    m_individually_accessible = individually_accessible;
    ...
endfunction
```

关键词：

uvm_reg，uvm_reg_filed，volatile

避坑指南：

在生成寄存器模型时，尽可能使用可靠的脚本。同时，对于某个域 volatile 的属性设定，也要在寄存器信息文件中标注。除了 IP-XACT 格式中有声明，如果采用的是公司内部的寄存器信息，那么在转换为标准寄存器信息文件时需要考虑 volatile 的属性补充；或在生成寄存器模型后，就这些域的 volatile 属性做更新，以使其符合寄存器访问和检查的要求。

阅读手记：

3.4.5 uvm_reg 的回调函数{pre, post}_{write, read}的用途是什么？

uvm_reg、uvm_reg_field 的这些回调函数（虚方法）需要通过类继承的方式后续填充，它们会与其他 uvm_reg_cbs 回调函数一起采取先后、递归的方式执行完。它们都是为了就某些特定的寄存器地址、路径、数据等信息进行分析判断，再修改这些属性，或修改最后的访问状态 status 继而令后续访问中断。

不建议单独就某个寄存器在寄存器模型文件中补充 *pre_write()* 等相关回调函数。较为合适的方法是，通过继承 uvm_reg 类和 uvm_reg_field 类，在子类中单独补充 pre_write() 等回调函数，再由顶层进行类型覆盖，继而影响后续寄存器模型的整体行为。

关键词：
uvm_reg, uvm_reg_filed, pre_write, post_write, pre_read, post_read

避坑指南：
使用这些回调函数，优先考虑对通用寄存器类别访问时的处理。例如，通过这些回调方法实现覆盖率的收集或某些特定事件的触发（识别到访问某些寄存器时触发相应事件）。对于特定寄存器的访问识别，还可以通过 uvm_reg_cbs 相关回调函数进行单独添加。

参考代码： uvm_reg.svh（UVM 源代码）

```
// uvm_reg.svh

task uvm_reg::do_write (uvm_reg_item rw)
  ...
  // PRE-WRITE CBS - FIELDS
  begin : pre_write_callbacks
    uvm_reg_data_t  msk;
    int lsb;

    foreach (m_fields[i]) begin
      uvm_reg_field_cb_iter cbs = new(m_fields[i]);
      uvm_reg_field f = m_fields[i];
      lsb = f.get_lsb_pos();
      msk = ((1<<f.get_n_bits())-1) << lsb;
      rw.value[0] = (value & msk) >> lsb;
      f.pre_write(rw);
      for (uvm_reg_cbs cb=cbs.first(); cb!=null; cb=cbs.next()) begin
        rw.element = f;
        rw.element_kind = UVM_FIELD;
        cb.pre_write(rw);
      end

      value = (value & ~msk) | (rw.value[0] << lsb);
    end
  end
  rw.element = this;
```

```
        rw.element_kind = UVM_REG;
        rw.value[0] = value;

        // PRE-WRITE CBS - REG
        pre_write(rw);
        for (uvm_reg_cbs cb=cbs.first(); cb!=null; cb=cbs.next())
           cb.pre_write(rw);

        if (rw.status != UVM_IS_OK) begin
          m_write_in_progress = 1'b0;

          XatomicX(0);

          return;
        end
        ...
    endtask
```

阅读手记：

3.4.6 与 uvm_reg_cbs 相关的回调函数的用处是什么？

uvm_reg_cbs 及其相关回调函数类所实现的回调方法，类似于在 uvm_reg、uvm_reg_filed 中内置的回调函数 *pre_write()*、*post_write()* 等。用户需要区别它们之间的调用顺序，以及更重要的它们的使用场景。用户可以定义继承于 uvm_reg_cbs 的回调函数类，并且填充它们的回调函数，再将其与对应的 uvm_reg、uvm_reg_field 实例进行绑定。

关键词：

uvm_reg，uvm_reg_field，uvm_reg_cbs，uvm_callbacks

避坑指南：

uvm_reg_cbs 及其相关回调函数类在定义其子类并例化后，需要将它们绑定到目标 uvm_reg、uvm_reg_field 实例。它们的操作范围比 uvm_reg、uvm_reg_field 预定义的 *pre_write()* 等回调函数更窄，而且最终实现回调函数调用的流程并不相同。

参考代码： uvm_reg_cbs.svh（UVM 源代码）

```
typedef uvm_callbacks#(uvm_reg, uvm_reg_cbs) uvm_reg_cb;
typedef uvm_callback_iter#(uvm_reg, uvm_reg_cbs) uvm_reg_cb_iter;
typedef uvm_callbacks#(uvm_reg_backdoor, uvm_reg_cbs) uvm_reg_bd_cb;
typedef uvm_callback_iter#(uvm_reg_backdoor, uvm_reg_cbs)
     uvm_reg_bd_cb_iter;
```

```
typedef uvm_callbacks#(uvm_mem, uvm_reg_cbs) uvm_mem_cb;
typedef uvm_callback_iter#(uvm_mem, uvm_reg_cbs) uvm_mem_cb_iter;
typedef uvm_callbacks#(uvm_reg_field, uvm_reg_cbs) uvm_reg_field_cb;
typedef uvm_callback_iter#(uvm_reg_field, uvm_reg_cbs)
        uvm_reg_field_cb_iter;
```

阅读手记：

3.4.7　adapter 中的 provides_responses 属性的作用是什么？

如果 adapter 连接的总线 driver 一侧有独立的 response item 返回（即 driver 通过 *item_done(rsp)* 或 *put_response(rsp)* 返回 response item），那么应当将 uvm_reg_adapter::provides_responses 设置为 1。

```
// uvm_reg_adapter.svh

// Variable: provides_responses
// Set this bit in extensions of this class if the bus driver provides
// separate response items.
bit provides_responses;
```

设定后，寄存器的 *write()*、*read()* 等操作，间接地通过 *uvm_reg_map::{do_bus_write, do_bus_read}* 中的逻辑来完成。每次的前门读写都会完整地经历 *uvm_reg_adapter::{reg2bus, bus2reg}*，区别仅在于，如果 *uvm_reg_adapter::provides_responses=1*，那么还会从总线 driver 一侧获得 response item，即真实的总线返回的数据事务。

```
// uvm_reg_map.svh

task uvm_reg_map::do_bus_write (uvm_reg_item rw,
                    uvm_sequencer_base sequencer,
                    uvm_reg_adapter adapter);
  ...
  if (adapter.provides_responses) begin
    uvm_sequence_item bus_rsp;
    uvm_access_e op;
    // TODO: need to test for right trans type, if not put back in q
    rw.parent.get_base_response(bus_rsp);
    adapter.bus2reg(bus_rsp,rw_access);
  end
  else begin
    adapter.bus2reg(bus_req,rw_access);
  end
```

```
        ...
      endtask
```

关键词：

uvm_reg_adapter，write，read，provides_responses，reg2bus，bus2reg

避坑指南：

provides_responses 属性的设定可以在实现 adapter 时添加。因为 adapter 绑定的总线 agent，其 driver 是否单独返回 response item 是固定的行为，所以该属性不需要在后续环境集成时修改。

```
class rkv_watchdog_reg_adapter extends uvm_reg_adapter;
  `uvm_object_utils(rkv_watchdog_reg_adapter)
  function new(string name = "rkv_watchdog_reg_adapter");
    super.new(name);
    provides_responses = 1;
  endfunction
  ...
endclass
```

阅读手记：

3.4.8 多个 uvm_reg_block 和 uvm_reg_map 的关系如何影响对寄存器的访问？

一个 uvm_reg_block 可以有一个或多个 uvm_reg_map。不同的 uvm_reg_map 可以设置总线位宽、寄存器的偏移地址等信息，与 uvm_reg_adapter、uvm_sequencer、uvm_reg_predictor 建立起联系。在支持前门访问时，不同的 uvm_reg_map 中有关寄存器的地址信息将通过 driver、interface 体现出来。

对于更复杂的层次化寄存器模型，顶层 uvm_reg_block 和 uvm_reg_map 将影响在顶层访问寄存器时采用的总线类型、地址信息等行为。与顶层 uvm_reg_map 有关的总线访问将通过 monitor、predictor、adapter 反映出整个 uvm_reg_block 的信息。要更准确地监测子一级环境中的寄存器访问，建议保留子一级验证环境中的 uvm_reg_block、uvm_reg_map 以及对应的 monitor、predictor、adapter 之间的关系。

关键词：

uvm_reg_block，uvm_reg_map，uvm_reg_adapter，uvm_reg_predictor

避坑指南：

每个 uvm_reg_block 可以有多个 uvm_reg_map，每个 uvm_reg_map 可以映射不同的地址

信息和总线。在复用模块级的测试用例时，其原先访问时采用的模块 uvm_reg_map 信息在顶层环境中，将会转而使用顶层的 uvm_reg_map 信息以及顶层的总线驱动。这种更替 uvm_reg_map 信息的方式有助于利用寄存器模型完成的读写序列的复用。

阅读手记：

3.4.9 如果并行利用 RGM 对寄存器做读/写可能出现什么问题？

不建议利用寄存器模型对某些寄存器做并行读/写（例如，使用 fork-join 语句实现并行读/写）。更准确地讲，不建议利用同一个 uvm_reg_map 对寄存器做出并行读/写行为。

多数情况下，对寄存器的读写是不需要这样做的。因为本质上它们最终都是通过同一个 uvm_reg_map 及其关联的 bus driver 做寄存器访问的。如果利用同一个 uvm_reg_map 做出寄存器的并行访问，那么时序可能会带来意料之外的紊乱。原因在于，在 uvm_reg_map 中有关的读/写实现逻辑中，当 *uvm_reg_adapter::provides_responses=1* 时，从 bus driver 获得 response item。如果有多个并行访问，那么由于缺少必要的 transaction_id 的识别，可能带来多个 request item 与 response item 之间不匹配的问题，继而影响寄存器读/写数据的返回。

关键词：
uvm_reg_block，uvm_reg_map，fork，provides_responses， response

避坑指南：
如果必须实现对某些寄存器的并行访问，那么建议绕过寄存器模型，直接在总线层利用 *get_response(output RSP response, input int transaction_id = -1)* 中的 transaction_id 参数完成每个 request item 的准确匹配。

阅读手记：

3.4.10 寄存器模型结构是否支持多个 top RGM？

uvm_reg_block 检查每个 top RGM（register model）的名字。因为 RGM 是 uvm_object 类型，它的 *get_full_name()* 方法并不会计算例化它的 component 层次，只会依据它所在 top RMG 的名字来计算它们名称的唯一性。从这个嵌套关系可以看到，每个 REG、RGM 的名字都是以它所属的 top RGM 为路径来计算的。

这也提醒我们，在例化 RGM 时，按照 *{get_full_name(), ".rgm"}* 是比较合理的一个方式。uvm_reg_block 提供了一个接口函数 *get_root_blocks()*，我们可以通过这个静态函数查找全局的 top RGM（存放于 *uvm_reg_block::m_roots* 中），然后通过它们的"名字"找到我们需要的 RGM。所以，UVM 可以支持多个 top RGM，但需要注意在例化时给它们提供不同的名称。

关键词：
uvm_reg_block，uvm_object，top，get_root_blocks

避坑指南：
尽管可以有多个 top RGM，但这种情况在验证环境中并不多。因为寄存器模型结构往往与设计结构一致，多数都是单顶层的树状结构。如果需要有多个 top RGM，那么要注意它们各自的子一级 uvm_reg_block 实例无法共享。这是因为 uvm_reg_block 在例化后需要调用 *uvm_reg_block::configure()* 函数来完成 RGM 树状结构的建立，而子一级的 uvm_reg_block 只能拥有一个 parent uvm_reg_block 实例。

参考代码： uvm_reg_blocks.svh（UVM 源代码）

```
// uvm_reg_block.svh
virtual class uvm_reg_block extends uvm_object;
  local uvm_reg_block   parent;
  local static bit      m_roots[uvm_reg_block];
  local int unsigned    blks[uvm_reg_block];
  local int unsigned    regs[uvm_reg];
  local int unsigned    vregs[uvm_vreg];
  local int unsigned    mems[uvm_mem];
  local bit             maps[uvm_reg_map];
  ...
endclass

function void uvm_reg_block::get_root_blocks(ref uvm_reg_block blks[$]);
  foreach (m_roots[blk]) begin
    blks.push_back(blk);
  end
endfunction: get_root_blocks
```

阅读手记：

3.4.11 uvm_reg_map 的数据位宽如果与总线不同需要做什么处理？

此处以 V3 验证课程中 DMAC 模块的验证环境为例。DMAC 的 RGM（register model）数据位宽是 64 位，但 DMAC 的 AHB slave 接口（和验证环境中的 AHB master 接口）位宽为

32 位。那么这就无法直接使用以往的（或 VIP 自身提供的）uvm_reg_adapter。这里需要将不同的位宽在 *uvm_reg_adapter::{reg2bus(), bus2reg()}* 函数中做额外的转换。

关键词：

uvm_reg_map，data width，uvm_reg_adapter

避坑指南：

如果寄存器模型中的 uvm_reg 持有 64 位数据位宽，且可能其高 32 位有实际用途，那么在这种情况下，就有必要将 64 位数据位宽转为 32 位总线的连续访问。如果相反，需要将 uvm_reg 的 32 位数据转为 64 位的数据总线访问，那么就需要考虑地址转换、字节使能等因素，以确保 32 位数据的有效读写，且避免其他地址的数据读写。

参考代码： rkv_dmac_reg2ahb_adapter.sv（V3 验证课程代码）

```
class  rkv_dmac_reg2ahb_adapter  #(type  T=svt_ahb_master_transaction)
extends lzn_common_reg2ahb_adapter #(T);
    `uvm_object_utils(rkv_dmac_reg2ahb_adapter)
    function new(string name="rkv_dmac_reg2ahb_adapter",
                  uvm_component parent=null);
      super.new(name);
    endfunction : new

    //need to define it as virtual function, usr maybe need to override it
    virtual function uvm_sequence_item reg2bus(
                                     const ref uvm_reg_bus_op rw);
      T trans = T::type_id::create("trans");;
      if(trans.get_class_name()=="svt_ahb_master_transaction") begin
         trans.cfg = master_cfg;
      end
      if(!(trans.randomize()with
          {burst_type == T::INCR; trans.data.size() == 2;}))
      begin
        uvm_report_warning("RNDFLD",
                          "Randomization failed in reg2bus() function");
      end
      trans.xact_type = (rw.kind ==UVM_READ)? T::READ:T::WRITE;
      trans.addr      = rw.addr;
      trans.burst_size= T::BURST_SIZE_32BIT;
      if(trans.xact_type == T::WRITE) begin
         trans.data[0]= rw.data[31:0];
         trans.data[1]= rw.data[63:32];
      end
      `uvm_info(get_name(), $sformatf("reg2bus rw::kind: %s, addr: %0x, data: %0x, status: %s", rw.kind, rw.addr, rw.data, rw.status), UVM_LOW)
      this.reg2bus_cb(rw);
      return trans;
    endfunction : reg2bus
```

```
    ...
    Endclass
```

阅读手记：

3.4.12 uvm_reg_map 的数据位宽如果与子一级不同需要做什么处理？

在系统集成过程中，模块级的 uvm_reg_map 的数据位宽可能与顶层的 uvm_reg_map 数据位宽不一致。这是因为模块级的数据访问总线宽度可能与顶层的数据访问总线宽度不一致，也可能是由不同设计的寄存器自身位宽的差异造成的。在这种情况下，就需要考虑不同位宽之间的适配处理，否则可能出现预料之外的寄存器读写行为。

在下面的集成方案中，顶层的 *uvm_reg_map ext_init_map* 是 32 位数据位宽，而其中的 *dmac.default_map* 又是 64 位数据位宽。顶层所关联的总线位宽是 32 位数据位宽，那么通过顶层 RGM ext_init_map 访问 dmac 寄存器时，就会存在数据位宽的转换问题。实际上，VCS 预置的 UVM RGM 源代码已经帮我们考虑到了这部分数据位宽转换问题，不需要额外添加代码。不过，请在编译时添加+*define+UVM_REG_BYTE_ADDRESSING_FIX*。

关键词：

uvm_reg_map，data_width，UVM_REG_BYTE_ADDRESSING_FIX

避坑指南：

这一解决办法是 VCS 预置的 UVM 源代码。这部分宏使能的代码用来解决不同层次 uvm_reg_map 数据位宽不一致时的位宽转换问题。不过，遗憾的一点是，原生的 uvm-1.2 版本并未解决这一问题。其他仿真器预置的 uvm-1.2 代码还需要在使用时确定是否有对这一问题的修复方案，希望 UVM 的下一个版本能够提供这个问题的修复方案。

参考代码： mcss_top_reg_model_block.sv（V3 验证课程代码），uvm_reg_map.svh（VCS 预置 UVM 代码）

```
// mcss_top_reg_model_block.sv
function void mcss_top_reg_model_block::create_address_maps();
`ifndef MCSS_TOP_RGM_USE_EMPTY
  ext_init_map = create_map("ext_init_map", 'h0, 4, UVM_LITTLE_ENDIAN);
  ext_init_map.add_submap(wdt0.map, MCSS_RGM_WDT0_BASE_ADDR);
  ext_init_map.add_submap(dualtimer0.map, MCSS_RGM_DUALTIMER0_BASE_ADDR);
  ext_init_map.add_submap(rtc0.map, MCSS_RGM_RTC0_BASE_ADDR);
  ext_init_map.add_submap(rtc1.map, MCSS_RGM_RTC1_BASE_ADDR);
  ext_init_map.add_submap(ssp0.map, MCSS_RGM_SSP0_BASE_ADDR);
  ext_init_map.add_submap(ssp1.map, MCSS_RGM_SSP1_BASE_ADDR);
  ext_init_map.add_submap(timer0.map, MCSS_RGM_TIMER0_BASE_ADDR);
```

```systemverilog
    ext_init_map.add_submap(timer1.map, MCSS_RGM_TIMER1_BASE_ADDR);
    ext_init_map.add_submap(trng.map, MCSS_RGM_TRNG_BASE_ADDR);
    ext_init_map.add_submap(uart0.map, MCSS_RGM_UART0_BASE_ADDR);
    ext_init_map.add_submap(uart1.map, MCSS_RGM_UART1_BASE_ADDR);
    ext_init_map.add_submap(i2c0.default_map, MCSS_RGM_I2C0_BASE_ADDR);
    ext_init_map.add_submap(i2c1.default_map, MCSS_RGM_I2C1_BASE_ADDR);
    ext_init_map.add_submap(gpio0.default_map, MCSS_RGM_GPIO0_BASE_ADDR);
    ext_init_map.add_submap(gpio1.default_map, MCSS_RGM_GPIO1_BASE_ADDR);
    ext_init_map.add_submap(gpio2.default_map, MCSS_RGM_GPIO2_BASE_ADDR);
    ext_init_map.add_submap(gpio3.default_map, MCSS_RGM_GPIO3_BASE_ADDR);
    ext_init_map.add_submap(dmac.default_map, MCSS_RGM_DMAC_BASE_ADDR);
    ext_init_map.add_submap(i2s0.default_map, MCSS_RGM_I2S0_BASE_ADDR);
`endif // mcss_top_RGM_USE_EMPTY
endfunction: create_address_maps

// uvm_reg_map.svh
function int uvm_reg_map::get_physical_addresses
              (uvm_reg_addr_t    base_addr,
               uvm_reg_addr_t    mem_offset,
               int unsigned      n_bytes,
               ref uvm_reg_addr_t addr[]
`ifdef UVM_REG_BYTE_ADDRESSING_FIX
               ,input bit upscaling);
`else
               );
...
    // Scale the consecutive local address in the system's granularity
`ifdef UVM_REG_BYTE_ADDRESSING_FIX
    if (m_byte_addressing)  begin
      //moving from byte based to byte based
      if (up_map.get_byte_addressing()==1) begin
        //Scale if going to wider  map
        if (bus_width < up_map.get_n_bytes(UVM_NO_HIER))
          k = (up_map.get_n_bytes(UVM_NO_HIER) -1)/bus_width + 1;
        else
          k = 1;
      end else begin //Moving from byte based to  non-byte based
        // if needs multiple physical address then don't scale.
        if ((local_addr.size > 1) && !(upscaling)) begin
          k = 1;
        end else begin
          //Scale if going to wider  map
          if (bus_width <= up_map.get_n_bytes(UVM_NO_HIER))
            k = 1.0/bus_width;
          else
            k = real'(up_map.get_n_bytes(UVM_NO_HIER))/bus_width;
        end
```

```
                end
            end else begin
                //moving from non-byte based to byte based
                if (up_map.get_byte_addressing()==1) begin
                    if (bus_width < up_map.get_n_bytes(UVM_NO_HIER))
                        k = up_map.get_n_bytes(UVM_NO_HIER);
                    else
                        k = bus_width;
                end else begin // From non-byte based to non-byte based
`endif
                    if (bus_width <= up_map.get_n_bytes(UVM_NO_HIER))
                        k = 1;
                    else
                        k = ((bus_width-1) / up_map.get_n_bytes(UVM_NO_HIER)) + 1;
`ifdef UVM_REG_BYTE_ADDRESSING_FIX
                end
            end
`endif
    ...
    Endfunction
```

阅读手记：

3.4.13 寄存器模型的镜像值和期望值什么情况下相等或不相等？

通过寄存器模型发起的 write 或 read 动作，如果是前门访问，那么在访问完成时，总线监测到的数据信息会通过 predictor 去标注反馈到寄存器模型中，继而一起更新镜像值（mirrored value）和期望值（desired value）；如果是后门访问，那么会在访问时一起更新镜像值和期望值。

这么来看，无论是镜像值还是期望值，在通过 predictor 或 *uvm_reg_map::m_auto_predict = 1* 时，都是为了与硬件实际值保持一致。对于寄存器的配置，也可以通过 *uvm_reg::set()* 和 *uvm_reg::upadte()* 的方式来完成。在这种情况下，set 动作会先更新期望值，而 update 动作则会触发一个实际的写动作。触发写动作的原因是这种方式会让期望值先于镜像值而变化，在写动作完成后，镜像值又会与期望值、硬件实际值保持一致。

对寄存器值的获取也可以通过 *uvm_reg::mirror()* 来完成（*uvm_reg_file::m_volatile = 1* 可触发每次的读操作）。在读操作完成后，镜像值和期望值也会同硬件实际值保持一致。

关键词：

mirrored value，desired value，predictor，m_auto_predict，set，update，mirror

避坑指南：

只有理解了 *update()* 和 *mirror()* 这两个方法触发写操作和读操作的条件，才不会因为这两个方法有时没有触发读写操作而困惑。当镜像值和期望值恢复一致后，通过 *uvm_reg::get()* 的动作获得的期望值也可以看做是镜像值（同硬件实际值）。另外，如果使用了 predictor 更新寄存器的期望值和镜像值，则无须使用 *uvm_reg_map::set_auto_predict()*。

阅读手记：

3.4.14 uvm_reg 的读/写动作在发起后没有结束的原因可能是什么？

uvm_reg::{write(), read()} 发起之后动作没有结束，已知目前的常见原因可能有以下几个。这里以 *write()* 动作为例，它在完成之前有 2 个阻塞动作。一个动作是 "*bus_req.end_event.wait_on();*"，一些工程师会发现代码在这句话被阻塞住。这其实是他们或者是在编译时添加了编译选项 "*+define+UVM_DISABLE_AUTO_ITEM_RECORDING*"（多数情况下是在使用 Synopsys VIP 时给的编译建议），或者是调用了 *uvm_sequencer::disable_auto_item_recording()* 函数，这些都会要求用户在 driver 一侧为实现事务记录（transaction recording）而手动调用 *uvm_sequence_item::{begin_tr()、accept_tr()、end_tr()}* 等函数。那么，如果添加了这个宏或函数又无法移除，为了让 RGM 与 driver 之间能够协调而不被阻塞，用户要在 driver 一侧发送数据后手动调用 *uvm_sequence_item::end_tr()* 函数，即能够间接触发 *uvm_sequence_item::end_event*，继而让 *uvm_reg_map::do_bus_write()* 任务不被此处语句所阻塞。

关键词：

multi-threaded，uvm_reg，write，read，provides_response

避坑指南：

uvm_reg_map::do_bus_write() 可能被阻塞的第二个地方在于，是否配置了 *uvm_reg_adapter::provides_responses = 1*。如果做了配置，就意味着要从 driver 一侧获得 response item，因此，driver 一侧需要将反馈数据通过诸如 *item_done(RSP)* 等方式反馈给 sequence。如果 driver 一侧没有反馈数据而 *uvm_reg_adapter::provides_responses = 1*，那么这将成为 write() 动作第二个会被阻塞的地方。

参考代码： uvm_reg_map.svh（UVM 源代码）

```
// uvm_reg_map.svh
task uvm_reg_map::do_bus_write (uvm_reg_item rw,
                                uvm_sequencer_base sequencer,
```

```
                                  uvm_reg_adapter adapter);
    ...
     // perform accesses
     foreach(accesses[i]) begin
       uvm_reg_bus_op rw_access=accesses[i];
       uvm_sequence_item bus_req;

       adapter.m_set_item(rw);
       bus_req = adapter.reg2bus(rw_access);
       adapter.m_set_item(null);

       if (bus_req == null)
         `uvm_fatal("RegMem",
          {"adapter [",adapter.get_name(),"] didnt return a bus transaction"});

       bus_req.set_sequencer(sequencer);
       rw.parent.start_item(bus_req,rw.prior);

       if (rw.parent != null && i == 0)
         rw.parent.mid_do(rw);

       rw.parent.finish_item(bus_req);
       bus_req.end_event.wait_on();

       if (adapter.provides_responses) begin
         uvm_sequence_item bus_rsp;
         uvm_access_e op;
         // TODO: need to test for right trans type, if not put back in q
         rw.parent.get_base_response(bus_rsp);
         adapter.bus2reg(bus_rsp,rw_access);
       end
       else begin
         adapter.bus2reg(bus_req,rw_access);
       end
       ...
     endtask
```

阅读手记：

第 4 章

Testbench 疑难点集合

4.1 编译与导入

设计模块与验证平台在仿真前要进行文件的编译与导入。文件命名规范、库和包的合理组织，这些都有助于对多个子系统验证环境和系统验证环境的有效管理。接下来我们将就文件的编译与导入讨论若干疑难点。

4.1.1 package 中可以定义什么类型？

package 中可以定义数据变量、方法（task、function）、DPI 方法声明、class、parameter、covergroup 和 property（断言属性）等。

关键词：

package，type

避坑指南：

package 中不可以包含（或编译）module 和 interface。package 中定义的方法、变量默认为静态类型，但建议在定义方法时将其声明为 automatic。

参考代码： tb_package_defs.sv

```
package rkv_pkg;
  parameter int ADDR_WIDTH = 16;
  parameter int DATA_WIDTH = 32;

  int cur_trans_id = 0;

  typedef enum {IDLE, RUN, WAIT, ABORT} state_t;
  state_t cur_trans_state = IDLE;

  class packet;
    rand byte unsigned data[];
  endclass

  covergroup data_cg with function sample(byte unsigned data);
```

```
        coverpoint data;
      endgroup
    endpackage

    module tb;
      import rkv_pkg::*;
      packet ps[];
      data_cg cg;
      initial begin
        ps = new[5];
        cg = new();
        foreach(ps[i]) begin
          packet p;
          ps[i] = new();
          p = ps[i];
          rkv_pkg::cur_trans_id = i;
          void'(p.randomize());
          foreach(p.data[j]) begin
            cg.sample(p.data[j]);
          end
        end
      end
    endmodule
```

阅读手记：

4.1.2 library 和 package 怎么区分？

library（库）是将以 module 和 interface 为主要的类型编译并归置到一起的"容器"。在链接过程（elaboration）中，允许使用 library 来解决 module 同名和不同设计组交付模块的问题。

package（包）更倾向于软件一侧的概念，而且它实际也只允许将数据类型或软件类型（如 class）置于其中，并进一步被编译到 library 中。

关键词：

library，package，compilation

避坑指南：

将软件类型都先置于 package 中，将硬件类型都置于 library 中。

阅读手记：

4.1.3 文件中出现 typedef class X 是什么意思？

这需要与常见的 *typedef enum*、*typedef struct* 的类型定义做区别。真正的类的定义依然是利用 "*X extends ...*" 的方式实现，而这里则是为了编译时能够让编译器识别 class X 而采取的"占坑"方法。即，先"声明"类 X 的存在，并且在后续对 X 的定义字段 "*X extends ...*" 中进行补充编译。

关键词：

typedef class，declaration，extends

避坑指南：

typedef class 常用来解决两个类之间的互相引用问题，即在 A 中使用 B 且在 B 中使用 A 的场景。

参考代码： tb_class_multi_reference.sv

```
package rkv_pkg;
  typedef class packet2;
  class packet1;
    packet2 pkt2;
  endclass
  class packet2;
    packet1 pkt1;
  endclass
endpackage

module tb;
  rkv_pkg::packet1 pkt1;
  rkv_pkg::packet2 pkt2;
  initial begin
    pkt1 = new();
    pkt2 = new();
  end
endmodule
```

阅读手记：

4.1.4 `include 和 import 的差别在哪里？

SystemVerilog 常会用`include 将多个文件"平铺"（flatten）置于某个域中（scope），这个域可能是 package、module、interface 等。简单理解，`include 就是将对应文本的内容"平铺"到当前域的字段中；import 则是从包（package）中引用某些需要的类型如 class、parameter、enum 到当前域中，以帮助编译器识别被引用的类型。

关键词：

`include，scope，import，package

避坑指南：

`include 部分会由编译器编译，import 部分会由编译器从库中查找导出。此外，对 import 的功能可能的误解在于，import 只是为了从其他库中导入某些类型，但这些导入的类型并不会再次被编译到其他域（如 package）中，这一点与 export 不同（但 export 使用的场合也非常有限）。

参考代码： tb_include_import.svh，tb_include_import.sv

```
// sv_include_import.svh
typedef enum {WRITE, READ, IDLE} rkv_op_t;

class rkv_packet;
  rand int unsigned data;
  rand rkv_op_t t;
endclass

class rkv_transaction;
  rand rkv_packet pkts[];
  constraint cstr {pkts.size inside {[3:5]};}
  function void post_randomize();
    foreach(pkts[i]) begin
      pkts[i] = new();
      void'(pkts[i].randomize());
    end
  endfunction
endclass

// sv_include_import.sv
package rkv_pkg;
  `include "sv_include_import.svh"
endpackage

package test_pkg;
  import rkv_pkg::rkv_transaction;
  class rkv_test;
    rand rkv_transaction trans[];
```

```
    constraint cstr {trans.size inside {[4:6]};}
    function void post_randomize();
      foreach(trans[i]) begin
        trans[i] = new();
        void'(trans[i].randomize());
      end
    endfunction
  endclass
endpackage

module tb;
  import rkv_pkg::rkv_transaction;
  import test_pkg::rkv_test;

  rkv_transaction tr;
  rkv_test test;

  initial begin
    test = new();
    void'(test.randomize());
    foreach(test.trans[i]) begin
      tr = test.trans[i];
      $display("test.trans[%0d].pkts dynamic array size is %0d",
               i, tr.pkts.size());
    end
    $finish();
  end
endmodule
```

仿真结果：

```
test.trans[0].pkts dynamic array size is 3
test.trans[1].pkts dynamic array size is 5
test.trans[2].pkts dynamic array size is 5
test.trans[3].pkts dynamic array size is 5
```

阅读手记：

4.1.5 `include 应该在哪里使用？

`include 的使用多见于在包（package）文件中将多个其他类文件"平铺"于其中，从而在编译时能够将多个文件中定义的类置于这个包中，形成一种逻辑上的包含关系。在 module

文件中也可能会使用 `include，使其"平铺"一些宏定义文件或接口文件。

关键词：

`include，package，module

避坑指南：

编译时一个 package 可以通过 `include 包含多个文件。该 package 被编译完成后，不再需要对被 `include 的文件做二次编译。

阅读手记：

4.1.6 应该怎么理解域（scope）呢？

SystemVerilog 的硬件类型 module、interface，或其软件类型 class、package，都涉及变量或方法调用时要考虑的域（scope）。简而言之，域是可以嵌套的，被嵌套的低一级的域可以"看到"高一级的域，不同域中的变量和方法可以同名，但各自的空间和生命周期则是独立的。

关键词：

scope，module，interface，class，package，variable，method

避坑指南：

在声明及引用变量和方法时，要注意它们所处的域；在引用 package 中的变量和方法时，还要使用 import。

参考代码：tb_scope_variable.sv

```
module tb;
  int a = 10;
  int b = 20;
  initial begin: init_proc
    int a = 30;
    int b = 40;
    $display("init_proc:: a = %0d, b = %0d", a, b);
    print_var();
  end
  initial begin
    $display("tb:: a = %0d, b = %0d", a, b);
  end
  function automatic print_var();
    int a = 50;
    int b = 60;
    $display("print_var:: a = %0d, b = %0d", a, b);
```

```
        endfunction
    endmodule
```

仿真结果：

```
    init_proc:: a = 30, b = 40
    print_var:: a = 50, b = 60
    tb:: a = 10, b = 20
```

阅读手记：

4.1.7 在系统验证阶段如何避免反复编译以节省时间？

各家仿真器在编译加速方面都在想办法。例如，VCS 的 partition compile、parallel compile，可以显著地缩短编译时间，使用时可以添加这样的链接命令 "*-partcomp -fastpartcomp=j4 +optconfigfile+partcomp.config*"。不过在系统测试阶段，如果仍然有较多活跃的 UVM test、sequence，将不可避免地带来测试用例的修改和反复编译。即便使用以上优化编译的手段，反复编译带来的额外开销仍然是惊人的，将影响效率。这里，我们建议尽可能地将 UVM 部分的 sequence 以任务的方式进行封装，再将任务（以及必要的参数）导出为 DPI-C 接口，继而在 C 一侧完成测试用例的移植。

关键词：

partition compile，parallel compile，uvm，dpi，tcl

避坑指南：

选择 UVM 一侧的哪部分 sequence 以及如何将其封装为任务（同时提供哪些参数），往往需要在经历过一次完整的系统测试后，将需求用尽可能简洁的任务和参数呈现出来，再将其导出为 DPI-C 接口。

```
    // partcomp.config file content
    partition package uvm_pkg svt_uvm_pkg svt_amba_common_uvm_pkg
svt_ahb_uvm_pkg svt_apb_uvm_pkg svt_axi_uvm_pkg svt_spi_uvm_pkg svt_uart_uvm_pkg
svt_i2s_uvm_pkg ;
    partition package lzn_common_pkg lzn_ahb_pkg lzn_apb_pkg lzn_cr_pkg
lvc_cr_pkg lzn_axi_pkg lzn_ioc_pkg lvc_ioc_pkg lvc_ioc2_pkg lzn_mem_pkg
lvc_jtag_pkg lzn_jtag_pkg;
    partition instance mcss_top_tb.dut ;
    partition package mcss_top_pkg ;
    partition package mcss_top_reg_model_pkg ;
    partition cell CortexM3Integration_sbst;
```

阅读手记：

4.1.8 如何解决和避免类重名或模块重名的问题？

类名、模块名等一般都应该带有某些固定前缀，比如所在子系统名、项目名、属类名等，这样就不容易出现重名的问题，而且也易于辨认该类型的身份（是不是为项目专属设计，是不是作为共用类型带有某些属类名等）。

如果类名发生重名，多数情况下可以通过 package 来实现类名的隔离。但是，如果要引用的 2 个 package 中有重名的部分，那么就往往不能采用 *import PACKAGE::** 这样的通配符去引用了，而是要单独指定引用的类型有哪些。

如果模块发生重名，为了后期编译的方便，建议修改 module 名称，这样可以不用为此而将重名的模块编入到不同 library 中，再由 configuration 指定引用关系。

关键词：

type name，class，module，package，library，rename，configuration

避坑指南：

不管是类名还是模块名，都可以考虑使用 DVT 这样的集成开发工具，使用它的重命名（rename）功能可协助重构（refactor）。这个功能可以帮助用户快速重命名任何有冲突的类型名称。如果你的工作环境缺少该工具，使用普通编辑器逐步手动修改也可以完成。不过，相比于事后修改，我们更建议团队在一开始工作时就规定好设计模块、验证组件等部分的命名规则。这样既能统一命名方式，也便于从名称中获取其用途、所在系统和项目等信息。

阅读手记：

4.2 验证组件实现

遵循常规的验证组件实现方式，可以快速构建起 UVM 验证平台。在实际工程中，对信号驱动、监测有更复杂的要求。用户需要考虑在组件内或序列内对某些信号做驱动和监测，而这不同于常规办法。验证结构在系统层面会展现更多的层次性、灵活性。如何控制不同的验证组件，并且让整个验证环境实现 phase 执行时的同步，是在实现验证组件和控制验证环境时遇到的疑难点。

4.2.1 监测器采样数据需要注意哪些？

监测器采样数据需要考虑以下几个方面：
（1）不是要对当前时钟驱动的信号做采样，而是对当前时钟"上一拍"的数据做采样。
（2）采样的数据要稳定。
（3）利用时钟块做采样，是在模拟触发器利用时钟沿做采样。
（4）有时我们在"时钟下降沿"做采样，而且没有利用时钟块，这其实是在模拟"组合逻辑"采样。即，在当前时钟周期内等待信号变化，待信号稳定后做采样。在时钟下降沿或时钟上升沿延迟一段时间（如 1ns）做信号采样都可以，但这时不再利用时钟块，因为使用时钟块一定会模拟时序采样（即在时钟上升沿采样上个周期的数据）。
（5）与采样数据分为"时序"采样和"组合"采样类似，驱动数据也是同样的做法。

关键词：
monitor，sample，stable，delay，clocking block，sequential，combinational

避坑指南：
采样时可以利用时钟块（clocking block）。但也要求在整个采样过程中恰当地运用时钟块，运用不当，有的地方采用了时钟块而有的地方没有使用时钟块，可能导致采样时序的逻辑错误。

阅读手记：

4.2.2 模块中的信号可以强制赋值和监测吗？

该问题的语境是指 module 中的信号（静态变量，static variable），而不包括 class 中的变量。在验证环境中，经常可以通过强制赋值（force）的方式修改某个层次下的信号，也可以通过层次化的方式直接获取某个层次下的信号。不过，为了保持验证环境与信号驱动、监测的独立性,我们建议将强制赋值、监测的方式封装在 interface 中。验证环境可以间接通过 interface 来满足这些需求。

关键词:
module,static,variable,force,monitor,interface

避坑指南:

模块中的动态变量(automatic variable)无法利用 SystemVerilog 的 force 完成变量值的修改。如果仅是调试目的,还可以在仿真过程中,利用仿真器的 value change(修改变量)功能修改某个时刻的动态变量值。

参考代码: tb_force_variable.sv

```
module mod;
  int data;
endmodule
module tb;
  class packet;
    int data;
  endclass
  mod m();
  initial begin: t1
    int addr;
    packet p = new();
    p.data = 10;
    // Error object's member cannot be forced
    // force p.data = 20;
    force m.data = 20;
    #20ns;
    $display("p.data = %0d, m.data = %0d, t1.addr = %0d",
             p.data, m.data, addr);
  end
  initial begin: t2
    #10ns;
    force t1.addr = 30;
  end
endmodule
```

仿真结果:

```
p.data = 10, m.data = 20, t1.addr = 30
```

阅读手记:

4.2.3 如何对设计层次中的某个实例进行侵入式赋值？

所谓的侵入式赋值（invasive assignment），就是对某个设计中的变量，在某个时间点进行驱动。驱动时要注意的是，这个外部测试平台提供的驱动是长期的侵入式驱动，还是只在该时间点驱动而后将驱动权仍然交还到设计逻辑。

这里以 VCS 的命令 force 为例，可以使用 *force <signal> <value>* 侵入式赋值，同时添加命令项 *-freeze*（表示该赋值将长期强制赋值）或 *-deposit*（表示其他驱动仍然可以为该 forced 变量进行赋值）。在系统验证中，我们有时需要将 VIP 的驱动连接到设计中，以此模拟设计中的驱动行为，同时忽视设计中原有的驱动，这时就需要实现侵入式赋值。这种赋值不同于 Tcl 中 force 命令的赋值，因为它要求时序或组合逻辑来模拟硬件行为。

关键词：

invasive，assign，drive，force，freeze，deposit

避坑指南：

侵入式赋值是通过 SystemVerilog 的 force 方式采取的强制赋值方式。在没有 release 语句的情况下，不会将驱动权交还给设计一侧的驱动逻辑。采用 SystemVerilog 的 force 语句使用时序或组合逻辑的驱动，比 Tcl 命令更灵活，也能够将测试平台中 VIP 的驱动逻辑通过接口和 force 赋值语句，对设计中任何层次的变量实现侵入式赋值。

参考代码： tb_design_force_invasivie_drive.sv

```
module rkv_mod (
  input   logic [3:0]   in_p0
  ,output logic [3:0]   out_p0
  );
  parameter string name = "rkv_mod";
  initial begin
    out_p0 <= 0;
    forever begin
      #5ns out_p0 <= out_p0 + 1;
      if(out_p0 == 'hF) break;
    end
  end
  always_comb $display("@%0t:: %s in_p0 = 'h%0x", $time, name, in_p0);
  always_comb $display("@%0t:: %s out_p0 = 'h%0x", $time, name, out_p0);
endmodule

module rkv_top;
  logic [3:0] w0, w1;
  rkv_mod #("m0") m0(w0, w1);
  rkv_mod #("m1") m1(w1, w0);
endmodule

interface rkv_intf;
```

```
// mimic driver from testbench
logic [3:0] d0, d1;
task drive(ref logic [3:0] d);
  d = 1;
  forever begin
    #5ns d = d << 1;
    if(d == 0) break;
  end
endtask
initial begin
  fork
    drive(d0);
    drive(d1);
  join
  $finish();
end
endinterface

module rkv_tb;
  rkv_top top();
  rkv_intf intf();
  // invasive force from testbench and seperate design drive logic
  always_comb begin
    force top.m0.out_p0 = intf.d0;
    force top.m1.out_p0 = intf.d1;
  end
endmodule
```

仿真结果：

```
@0::    m0 out_p0 = 'h1
@0::    m1 in_p0  = 'h1
@5000:: m0 out_p0 = 'h2
@5000:: m1 in_p0  = 'h2
@10000:: m0 out_p0 = 'h4
@10000:: m1 in_p0  = 'h4
@15000:: m0 out_p0 = 'h8
@15000:: m1 in_p0  = 'h8
```

阅读手记：

4.2.4 如何在仿真结束时打印一些测试总结信息？

对于实现数据比对的总结信息，可以在模块验证环境中的 scoreboard 中定义一些变量，以此统计数据总量（进、出）、比较次数、成功次数、失败次数等，继而在 report_phase() 中将这些数据信息进行格式化的打印。如果到了子系统环境，那么不同模块对应的独立的 scoreboard 也能够起到监测、比对和总结作用，不需要为它们的工作担心。

在仿真结束时可以打印测试是否通过的信息。这有两个参考标准，一个标准是根据 UVM_ERROR 数量来统计测试是否通过，一个标准是根据 scoreboard 数据比对的信息数量来统计测试是否通过。前者的统计更加严格，后者的统计只以数据比对为准。在以 UVM_ERROR 数量为参考时，可能要降低某些情况写 UVM_ERROR 的消息等级，或干脆屏蔽这些信息（修改 uvm_action）以避免它们的报错显示。在仿真结束时，可以利用 report server，根据 UVM_ERROR、UVM_FATAL 数量判断测试是否通过：

```
uvm_report_server rps = uvm_report_server::get_server();
if(rps.get_severity_count(UVM_ERROR) == 0) test_status = TEST_PASS;
else test_status = TEST_FAILED;
```

关键词：

scoreboard，report_phase，compare，UVM_ERROR，uvm_report_server

避坑指南：

如果在仿真最后打印测试是否通过的信息，那么还要考虑修改回归测试工具。比如，考虑它是按照哪种分析方式来查看测试是否通过，是根据 UVM 最后的报告总结信息，还是根据在 report_phase 中打印的 TEST_PASS、TEST_FAIL 的信息。

参考代码： tb_report_server_summary.sv

```
package rkv_pkg;
  import uvm_pkg::*;
  `include "uvm_macros.svh"
  class comp1 extends uvm_component;
    `uvm_component_utils(comp1)
    function new(string name = "comp1", uvm_component parent = null);
      super.new(name, parent);
    endfunction
    task reset_phase(uvm_phase phase);
      `uvm_info("PHASE", "reset phase entered at", UVM_LOW)
    endtask
    task main_phase(uvm_phase phase);
      `uvm_info("PHASE", "main phase entered", UVM_LOW)
      phase.raise_objection(this);
      #30us;
      `uvm_error("CMPDATA", "compare data failed")
      phase.drop_objection(this);
    endtask
  endclass
```

```systemverilog
class report_server_summary_test extends uvm_test;
  comp1 c1;
  `uvm_component_utils(report_server_summary_test)
  function new(string name = "report_server_summary_test",
               uvm_component parent = null);
    super.new(name, parent);
  endfunction
  function void build_phase(uvm_phase phase);
    c1 = comp1::type_id::create("c1", this);
  endfunction
  task reset_phase(uvm_phase phase);
    `uvm_info("PHASE",
              $sformatf("reset phase entered at %0t" , $time), UVM_LOW)
  endtask
  task main_phase(uvm_phase phase);
    `uvm_info("PHASE",
              $sformatf("main phase entered at %0t" , $time), UVM_LOW)
  endtask
  function void end_of_elaboration_phase(uvm_phase phase);
    uvm_report_server rps = uvm_report_server::get_server();
    uvm_default_report_server drps;
    if(!$cast(drps, rps)) `uvm_error("CASTFAIL", "TYPE CASTING ERROR")
    drps.enable_report_id_count_summary = 0;
  endfunction
  function void report_phase(uvm_phase phase);
    uvm_report_server rps = uvm_report_server::get_server();
    `uvm_info("REPORT",
              $sformatf("INFO COUNT: %0d",
                        rps.get_severity_count(UVM_INFO)),
              UVM_LOW)
    `uvm_info("REPORT",
              $sformatf("ERROR COUNT: %0d",
                        rps.get_severity_count(UVM_ERROR)),
              UVM_LOW)
    `uvm_info("REPORT",
              $sformatf("Test finished at %0t", $time), UVM_LOW)
  endfunction
endclass
endpackage

module tb;
  import uvm_pkg::*;
  `include "uvm_macros.svh"
  import rkv_pkg::*;
  initial run_test("report_server_summary_test");
endmodule
```

仿真结果：

```
    UVM_INFO @ 0: reporter [RNTST] Running test report_server_summary_test...
    UVM_INFO tb_report_server_summary.sv(10) @ 0: uvm_test_top.c1 [PHASE] reset phase entered at
    UVM_INFO tb_report_server_summary.sv(31) @ 0: uvm_test_top [PHASE] reset phase entered at 0
    UVM_INFO tb_report_server_summary.sv(13) @ 0: uvm_test_top.c1 [PHASE] main phase entered
    UVM_INFO tb_report_server_summary.sv(34) @ 0: uvm_test_top [PHASE] main phase entered at 0
    UVM_ERROR tb_report_server_summary.sv(16) @ 30000000: uvm_test_top.c1 [CMPDATA] compare data failed
    UVM_INFO tb_report_server_summary.sv(44) @ 30000000: uvm_test_top [REPORT] INFO COUNT: 6
    UVM_INFO tb_report_server_summary.sv(45) @ 30000000: uvm_test_top [REPORT] ERROR COUNT: 1
    UVM_INFO tb_report_server_summary.sv(46) @ 30000000: uvm_test_top [REPORT] Test finished at 30000000
```

阅读手记：

4.2.5 为什么有时无法在 sequence 或 test 中使用 force 语句？

使用 uvm_hdl_force 就没有这样的困惑了，因为它不需要将 sequences、tests 放置于顶层的 testbench 下方（或与 TB 文件一起在 $unit 域编译）。

如果确实需要使用 SystemVerilog force，可以将 sequences、tests 文件打包到一个头文件 .svh 中，并将它包裹在顶层 testbench 中，即可以给 sequences、tests 文件打开顶层 tb 以下层次的"视野"，继而访问层次化的信号。仿真器不支持将层次化索引的方法封装在 package 中，所以我们要将有关的层次化访问文件"裸露"出来。

关键词：

force，include，hdl_xmrl_force，uvm_hdl_force，hierarchical access

避坑指南：

示例中的头文件只是用来收集 sequences、tests，并不是将它们再封装到某一个 package 中，否则它们仍然"看不到"TB 以下的层次信号。

参考代码： tb_force_hdl_test.sv，tb_force_hdl_tb.sv

```
// tb_force_hdl_test.sv
class rkv_sequence extends uvm_sequence;
```

```systemverilog
  `uvm_object_utils(rkv_sequence)
  function new(string name = "rkv_sequence");
    super.new(name);
  endfunction
  task body();
    force tb.m.data1 = 10; // SV force
  endtask
endclass
class force_hdl_test extends uvm_test;
  `uvm_component_utils(force_hdl_test)
  function new(string name = "force_hdl_test",
               uvm_component parent = null);
    super.new(name, parent);
  endfunction
  task run_phase(uvm_phase phase);
    rkv_sequence seq = rkv_sequence::type_id::create("seq", this);
    phase.raise_objection(this);
    super.run_phase(phase);
    seq.start(null);
    force tb.m.data2 = 11; // SV force
    $hdl_xmr_force("tb.m.addr", "20"); // VCS supplied system function
    uvm_hdl_force("tb.m.cmd", 30);
    #10ns;
    phase.drop_objection(this);
  endtask
endclass

// tb_force_hdl_tb.sv
module mod;
  int data1;
  int data2;
  int addr;
  int cmd;
endmodule
module tb;
  import uvm_pkg::*;
  `include "uvm_macros.svh"
  `include "uvm_force_hdl_test.sv"
  mod m();
  initial run_test("force_hdl_test");
  initial
    #1ns $display("m.data1 = %0d, m.data2 = %0d, m.addr = %0d, m.cmd = %0d",
                  m.data1, m.data2, m.addr, m.cmd);
  int cmd;
  initial $hdl_xmr("tb.m.cmd", "cmd");
endmodule
```

仿真结果：

```
    m.data1 = 10, m.data2 = 11, m.addr = 20, m.cmd = 30
```

阅读手记：

4.2.6 为什么寄存器模型机构应与验证环境层次保持一致？

在多数验证环境下，我们期望有一个唯一的顶层寄存器模型（uvm_reg_block）实例。如果是模块验证环境，那么该建议是容易实现的；如果是系统级环境且嵌入多个子一级验证的环境，那么可能会有系统环境与子一级验证环境的寄存器模型之间发生冲突的情况发生。

寄存器模型的例化建议在 uvm_test 层进行并完成自顶向下的配置，以便将子一级寄存器模型句柄传到子一级验证环境中，继而保持整个环境具备唯一的顶层寄存器模型实例。因此，寄存器模型结构与设计层次、验证环境结构保持一致的原因，与配置对象（config object）结构一致性的原因类似，都是为了便于验证环境的管理。

关键词：
uvm_reg_block，singleton，configuration，environment，hierarchy

避坑指南：
如果有多个验证环境的嵌套，那么应该在 uvm_test 层例化唯一的顶层寄存器模型实例，并将其句柄传递到环境（uvm_env）结构中。在 uvm_env 中，进一步将子一级寄存器模型句柄传到子一级验证环境，以此方式实现递进式的寄存器模型传递。需要避免出现上级环境与下级环境的寄存器模型各自独立、缺少层次关联的情况，除非有特殊的使用目的。

阅读手记：

4.2.7 为什么建议执行任务时各组件统一使用 run_phase 或 main_phase？

如果只考虑 run_phase 和 main_phase 在各自的 domain 中顺序执行，而且都几乎会"同时"开始运行，那么不同验证组件之间不会出现执行 phase 的同步问题。但是，如果有 phase 跳转的需求，那么这两个 phase 任务在跳转时是互相独立的，它们之间不能跳转也不会统一跳转至某个 phase。

在复用底层 UVM 组件或环境时，如果它们采用的 run_phase 或 main_phase 与上层环境结构使用了不同的 phase 方法，那么对于 reset 事件的处理需要各自独立进行。对于 main_phase，可以采用 phase jump 的方式实现 reset_phase 和 main_phase 之间的跳转；对于 run_phase，请避免使用 phase jump，这是因为 uvm_root 只允许 run_phase 在 0 时刻运行，而运行后的 run_hase 如果发生 jump back，再次进入 run_phase 时则会违反规定。所以，run_phase 需要在自己内部通过监听 reset 事件的方式做好响应处理。

关键词：
run_phase，main_phase，reset，jump

避坑指南：
在实现验证环境时，避免自己的环境发生多个 run_phase 或 main_phase 并行出现的情况，否则会给组件之间 phase 的同步带来麻烦。VIP 等底层 UVC 组件多使用 run_phase，这种单一任务 phase 便于管理，也易于与上层环境兼容。在上层环境中，建议对 scoreboard、subscriber、coverage model、reference model 等组件采用 run_phase。可以将激励场景实现部分（sequence）采用 main_phase 方式管理，或统一使用 run_phase（方案较为整洁）。

参考代码： uvm_root.svh（UVM 源代码），tb_phase_individual_run_main_jump.sv

```
// uvm_root.svh
// It is required that the run phase start at simulation time 0
// TBD this looks wrong - taking advantage of uvm_root not doing anything else?
// TBD move to phase_started callback?
task uvm_root::run_phase (uvm_phase phase);
  // check that the commandline are took effect
  foreach(m_uvm_applied_cl_action[idx])
    if(m_uvm_applied_cl_action[idx].used==0) begin
      `uvm_warning("INVLCMDARGS",$sformatf("\"+uvm_set_action=%s\" never took effect due to a mismatching component pattern",m_uvm_applied_cl_action[idx].arg))
    end
  foreach(m_uvm_applied_cl_sev[idx])
    if(m_uvm_applied_cl_sev[idx].used==0) begin
      `uvm_warning("INVLCMDARGS",$sformatf("\"+uvm_set_severity=%s\" never took effect due to a mismatching component pattern",m_uvm_applied_cl_sev[idx].arg))
    end

  if($time > 0)
    `uvm_fatal("RUNPHSTIME",
      {"The run phase must start at time 0, current time is ",
       $sformatf("%0t", $realtime),
       ". No non-zero delays are allowed before ",
       "run_test(), and pre-run user defined phases may not consume ",
       "simulation time before the start of the run phase."})
endtask
```

```systemverilog
// tb_phase_individual_run_main_jump.sv
module tb;
  import uvm_pkg::*;
  `include "uvm_macros.svh"

  class phase_jump_test extends uvm_test;
    bit has_main_phase_jump;
    bit has_run_phase_jump;
    `uvm_component_utils(phase_jump_test)
    function new(string name = "phase_jump_test",
                 uvm_component parent = null);
      super.new(name, parent);
    endfunction
    function void start_of_simulation_phase(uvm_phase phase);
      `uvm_info("START_OF_SIMULATION", "phase running", UVM_LOW)
    endfunction
    task reset_phase(uvm_phase phase);
      phase.raise_objection(this);
      `uvm_info("RESET", "phase running", UVM_LOW)
      repeat(1) #1ns
        `uvm_info("RESET",
                  $sformatf("@%0t phase stepping..", $time), UVM_LOW)
      phase.drop_objection(this);
    endtask
    task main_phase(uvm_phase phase);
      phase.raise_objection(this);
      `uvm_info("MAIN", "phase running", UVM_LOW)
      repeat(2) #1ns
        `uvm_info("MAIN",
                  $sformatf("@%0t phase stepping..", $time), UVM_LOW)
      if(!has_main_phase_jump) begin
        has_main_phase_jump = 1;
        phase.jump(uvm_reset_phase::get());
      end
      phase.drop_objection(this);
    endtask
    task run_phase(uvm_phase phase);
      phase.raise_objection(this);
      `uvm_info("RUN", "phase running", UVM_LOW)
      repeat(3) #1ns
        `uvm_info("RUN",
                  $sformatf("@%0t phase stepping..", $time), UVM_LOW)
      if(!has_run_phase_jump) begin
        has_run_phase_jump = 1;
        phase.jump(uvm_start_of_simulation_phase::get());
      end
```

```
          phase.jump(uvm_start_of_simulation_phase::get());
          phase.drop_objection(this);
        endtask
      endclass
      initial run_test("phase_jump_test");
    endmodule
```

仿真结果:

```
      UVM_INFO @ 0: reporter [RNTST] Running test phase_jump_test...
      UVM_INFO tb_phase_individual_run_main_jump.sv(13) @ 0: uvm_test_top
[START_OF_SIMULATION] phase running
      UVM_INFO tb_phase_individual_run_main_jump.sv(33) @ 0: uvm_test_top [RUN]
phase running
      UVM_INFO tb_phase_individual_run_main_jump.sv(17) @ 0: uvm_test_top
[RESET] phase running
      UVM_INFO tb_phase_individual_run_main_jump.sv(34) @ 1000: uvm_test_top
[RUN] @1000 phase stepping..
      UVM_INFO tb_phase_individual_run_main_jump.sv(18) @ 1000: uvm_test_top
[RESET] @1000 phase stepping..
      UVM_INFO tb_phase_individual_run_main_jump.sv(23) @ 1000: uvm_test_top
[MAIN] phase running
      UVM_INFO tb_phase_individual_run_main_jump.sv(34) @ 2000: uvm_test_top
[RUN] @2000 phase stepping..
      UVM_INFO tb_phase_individual_run_main_jump.sv(24) @ 2000: uvm_test_top
[MAIN] @2000 phase stepping..
      UVM_INFO tb_phase_individual_run_main_jump.sv(34) @ 3000: uvm_test_top
[RUN] @3000 phase stepping..
      UVM_INFO tb_phase_individual_run_main_jump.sv(24) @ 3000: uvm_test_top
[MAIN] @3000 phase stepping..
      UVM_INFO ${UVM_HOME}/base/uvm_phase.svh(1556) @ 3000: reporter [PH_JUMP]
phase run (schedule , domain common) is jumping to phase start_of_simulation
      UVM_INFO ${UVM_HOME}/base/uvm_phase.svh(1556) @ 3000: reporter [PH_JUMP]
phase main (schedule uvm_sched, domain uvm) is jumping to phase reset
      UVM_INFO ${UVM_HOME}/base/uvm_objection.svh(1276) @ 3000: reporter
[TEST_DONE] 'run' phase is ready to proceed to the 'extract' phase
      UVM_INFO phase_individual_run_main_jump.sv(13) @ 3000: uvm_test_top
[START_OF_SIMULATION] phase running
      UVM_INFO phase_individual_run_main_jump.sv(33) @ 3000: uvm_test_top [RUN]
phase running
      UVM_FATAL ${UVM_HOME}/uvm_root.svh(1077) @ 3000: reporter [RUNPHSTIME]
The run phase must start at time 0, current time is 3000. No non-zero delays are
allowed before run_test(), and pre-run user defined phases may not consume
simulation time before the start of the run phase.
```

阅读手记：

4.2.8 如何更新 driver 的驱动行为？

一些商业 VIP 的组件，包括 driver，往往会在其驱动逻辑的各个关键时间点上预留若干回调函数的入口（多以 uvm_callback 实现）。这种方式便于用户后期修改 driver 驱动逻辑时，可以通过继承 uvm_callback 类、实现对应方法、再以回调函数对象绑定的方式来修改 driver 的驱动行为。

关键词：

driver，override，uvm_callback

避坑指南：

如果预留的回调函数无法满足对 driver 驱动行为的特定修改，那么还可以考虑定义一个继承于该 driver 的子类，重新实现目标方法。待完成后，通过类型覆盖或实例覆盖，用新的子类替换原有 driver 类型。不过要注意的是，重新实现的方法在父类中应该声明为虚方法，且在子类中尽可能避免声明新的成员（原有父类的代码并没有调用子类这些新成员的逻辑）。

参考代码： V3 验证课程代码

```
class lvc_i2c_master_driver_callback extends uvm_callback;
  `uvm_object_utils(lvc_i2c_master_driver_callback)
  function new(string name = "lvc_i2c_master_driver_callback");
    super.new(name);
  endfunction
  //-------------------------------------------------------------
  // define virtual methods below which later to be executed in
  // lvc_i2c_master_driver
  //-------------------------------------------------------------
  virtual function void post_seq_item_get(
                      lvc_i2c_master_driver driver,
                      lvc_i2c_master_transaction xact,
                      ref bit drop);
  endfunction
endclass

class lvc_i2c_master_driver extends
      uvm_driver #(lvc_i2c_master_transaction);
  ...
  `uvm_register_cb(lvc_i2c_master_driver,
                  lvc_i2c_master_driver_callback)
```

```
      ...
      task lvc_i2c_master_driver::post_seq_item_get_cb_exec(
            lvc_i2c_master_transaction xact, ref bit drop);
        post_seq_item_get(xact, drop);
        `uvm_do_callbacks(lvc_i2c_master_driver,
                          lvc_i2c_master_driver_callback,
                          post_seq_item_get(this, xact, drop))
      endtask
      ...
      task lvc_i2c_master_driver::consume_from_seq_item_port();
        lvc_i2c_master_transaction trans;
        lvc_i2c_master_transaction trans_out;
        bit success = 0;
        bit drop = 0;
        forever begin
          trans = new();
          `uvm_info("consume_from_seq_item_port",
                    "Get request item from sequencer...", UVM_DEBUG)
          seq_item_port.get_next_item(trans);

          `uvm_info("consume_from_seq_item_port",
                    "Got request item from sequencer...", UVM_DEBUG)
          // branches open here
          drop = 0;
          // callback called for loading exceptions from callbacks
          post_seq_item_get_cb_exec(trans, drop);
          if(drop) begin
            `uvm_info("consume_from_seq_item_port", "Drop bit set through
post_seq_item_get callback. Transaction dropped", UVM_DEBUG)
            // put response to seq_item_port
            seq_item_port.put_response(trans);
          end
          ...
      endtask
    endclass
```

阅读手记：

4.2.9 force 和 $hdl_xmr_force、uvm_hdl_force 等命令有什么差别？

首先，force 是 SystemVerilog 的强制赋值方式。只要能够看到 DUT 的层次信号，就可以做强制赋值。只是有些情况下，如果需要在 sequence、test 中使用 force，那么它们所处的文件还应该在 TB 中采用 `include 包裹的方式进行编译。

$hdl_xmr_force() 是仿真工具 VCS 提供的 force 命令（其他仿真器也提供类似的 force 命令），并非 SystemVerilog 原生的系统函数。类似的还有 *$hdl_xmr()*。*$hdl_xmr_force()*，可以通过指定信号所在的位置，对信号做强制赋值。

uvm_hdl_force() 则是 UVM 提供的 force 命令，通过 DPI 的方式而不依赖特定的仿真器，即能够"指哪打哪"，实现跨平台的强制赋值需求。

关键词：
force，$hdl_xmr_force，uvm_hdl_force

避坑指南：
要达到跨平台复用，推荐采用 *uvm_hdl_force()*。而且，UVM 的 package 编译也可以独立完成，不需要像使用 force 语句时那样，将所在文件通过 `include 被编译到 TB 顶层。

阅读手记：

4.3 测试平台控制

越是朝着系统级验证环境发展，对整个测试平台的控制要求越复杂。这种复杂性要求体现在几个方面。其一，需要考虑测试平台重复编译的成本，转而采用 DPI-C 接口来实现 UVM 测试用例到 C 测试用例的迁移；其二，验证环境中各个组件的驱动、监测、比较功能在不同测试用例中可能会有不同的控制，需要考虑不同测试用例中的各验证组件协调工作的问题；其三，为了更有效地利用仿真运算资源，如何降低仿真负载、减少非必要测试用例、对测试进行局部修改避免重新仿真等都问题也都需要考虑。接下来将围绕以上有关测试平台控制的疑难点展开讨论。

4.3.1 如何将覆盖率数据信息与测试用例关联？

这里以 VCS 仿真器为例。VCS 的覆盖率数据在合并时，默认不会将覆盖率数据的贡献与各测试用例做映射。但很多时候，我们需要分析是哪些测试用例对目标覆盖率做了贡献，那么在使用覆盖率数据命令 urg 时，添加选项 "urg -dir ... -show tests" 可以将各项覆盖率信息与测试用例做好映射关系。

关键词：

coverage，test，contribution，urg

避坑指南：

一般情况下，我们不需要单独关心某些覆盖率与测试用例之间的映射关系。但对于一些特定的覆盖率，能够看到是不是目标测试用例对覆盖率有实际的贡献，这有助于增强我们的测试用例和某些设计状态变化的信息感知能力。为了配合特定测试用例的收集，也可以只在特定测试用例过程中收集某些覆盖率（例如，使能某些 covergroup 的采样），这同样有利于覆盖率的定向收集。

阅读手记：

4.3.2 系统验证如何实现 C 用例和 UVM 用例之间的互动？

这里以 C 用例由硬核（hard core，即真实处理器，与 virtual core 虚核相对）执行，与 UVM 用例（sequence）之间的互动加以说明。由于硬核的参与，它与验证环境一侧的互动（事件同步、共享存储访问等）往往需要通过公共的存储空间，这部分空间可以采用片内存储或片外存储的形式，也可以采用某些设计模块闲置的寄存器地址来做交互。在这一基础上，封装一些用来同步的接口函数，就能够实现 C 用例与 UVM 用例之间的互动。

关键词：

C，UVM，sequence，hardcore，synchronization，memory，shared

避坑指南：

SoC 系统中往往有多个子系统，而各个子系统之间数十个处理器的 C 用例与 UVM 用例之间要实现同步，要考虑公共的存储空间是否可被所有硬核同时访问（这一方案相对而言简单、统一），或针对不同子系统的访问限制，需要考虑采取分布式的存储空间，实现子系统处理器的 C 用例和 UVM 测试用例之间的同步。

在实际项目中，建议 DFV（Design For Verification）在早期沟通中开拓部分可共用的存储地址空间、寄存器地址空间。这些空间几乎不会对设计带来额外的影响，更不会产生安全危害，但可以在系统测试阶段和片后测试阶段给验证带来更多的便利。

阅读手记：

4.3.3 系统验证时测试用例有误，是否可以避免重新仿真而只做局部修改？

系统验证阶段，不考虑因为测试用例的错误而需要重新编译测试平台和测试用例的耗费，重新执行仿真带来的额外消耗（以小时计）则需要考虑在内。测试用例的试错不单单在于测试用例本身，往往还与当前系统的稳定性、集成度以及对系统功能的理解程度相关。

在已经发现测试用例有误，或对测试用例正确性没有足够信心的情况下，可以考虑仿真器的 checkpoint 功能。利用仿真脚本在若干仿真节点保存现场，而后在最终仿真结果出错时，经过分析可以将仿真退回到之前保存好的某个现场（checkpoint）。在退回到特定现场之后，正常情况下仿真还会按照原有"剧情"发展，但我们可以通过 Tcl 命令（如强制赋值某些信号、存储内容或寄存器值）去影响测试的发展方向。

不管是修改测试意图还是修改原本有错的局部设计逻辑，都是为了降低试错成本，用更短的时间去印证自己对设计的理解、对测试的修改是正确的。在测试通过后，可以提交设计修改申请、重新编译测试平台或修改测试用例。

关键词：

system，recompilation，testcase，checkpoint，tcl，force

避坑指南：

以 VCS 为例，Tcl 命令不仅可以强制修改（force）部分信号、变量，还可以调用测试平台层次中某些 module、interface 中的函数。这可以丰富 Tcl 的可执行能力，让它实现更多的测试可能性。在系统验证阶段，充分利用 Tcl 的功能可以节省不少时间，加速测试用例的调试和完善。

参考代码： V3 验证课程代码

```
// Tcl part, call SV part method
# go to specific scope to call method it owns
scope mcss_top_tb.top_if
# call UVM debug function defined in the specific scope
# call memory access method to validate the memory content read & write
set rdata [call {mem_backdoor_readw(0x31000) }]
format 0x%x $rdata
call {mem_backdoor_writew(0x31000, 0x55FF33CC) }
set rdata [call {mem_backdoor_readw(0x31000) }]
format 0x%x $rdata

// SV part, mcss_top_if interface partial content
...

function void mem_backdoor_writew(int unsigned addr, int unsigned data);
  if(addr >= `MCSS_INTMEM_BASE_ADDR)
    intmem_backdoor_writew(addr, data);
  else
    extmem_backdoor_writew(addr, data);
endfunction

function int unsigned mem_backdoor_readw(int unsigned addr);
  if(addr >= `MCSS_INTMEM_BASE_ADDR)
    return intmem_backdoor_readw(addr);
  else
    return extmem_backdoor_readw(addr);
endfunction
```

阅读手记：

4.3.4 如何在仿真过程中更好地控制验证环境的结构和行为？

由于 UVM 结构是在仿真开始后动态构建的，所以相较于 Verilog 的静态结构和层次，UVM 结构可以通过配置的方法，在不同的测试中按需选择使能或关闭部分子环境或组件。

对 UVM 验证环境结构的控制可以写在 uvm_test 层次中（需要通过编译来实现），也可以使用 SystemVerilog 支持的系统命令（*$value$plusargs()*、*$test$plusargs()*）。接收系统命令的方式适合点状的命令传递，uvm_test 层往往无法直接确定需要的配置内容，这就要求测试在运行前将参数传进来。

关键词：

configuration，runtime，$value$plusargs()，$test$plusargs()

避坑指南：

如果目标设计的参数配置组合较多，那么除了 UVM 测试端本身去遍历这些组合，还可以考虑将这些对应于设计的参数外置为仿真时传入的运行参数。那样的话，这些配置参数将会是一个矩阵式的搭配。它们作为一个配置文件，选择不同格式传入到 uvm_test 中，再由其适配的解析函数去获取配置文件中的有关参数。这种方式可以更灵活地利用已有的测试激励 pattern，结合外部传入的多种参数组合，更好地控制验证环境结构和行为。

参考代码： V3 验证课程代码

```
// Makefile side to pass runtime options
ifneq ($(SEQUENCE),)
RUN_OPTS += +SEQUENCE_OVERRIDE=$(SEQUENCE)
endif
ifeq ($(BOOTMEM),EXTMEM)
RUN_OPTS += +CODE_EXTMEM_PATH=$(CODE_MEM_PATH)
else
RUN_OPTS += +CODE_INTMEM_PATH=$(CODE_MEM_PATH)
endif

// UVM base test side
// get simulation run options received
if($value$plusargs("SEQUENCE_OVERRIDE=%s", cfg.sequence_override)) begin
  set_type_override("mcss_top_base_virtual_sequence",
                    cfg.sequence_ override);
end
if($value$plusargs("CODE_EXTMEM_PATH=%s", cfg.code_extmem_path)) begin
  `uvm_info(get_type_name(),
            $sformatf("Get core code memory HEX file path %s",
                      cfg.code_extmem_path),
            UVM_LOW)
end
if($test$plusargs("HCORE_AVAIL")) cfg.hcore_avail = 1;
if($test$plusargs("VCORE_AVAIL")) cfg.vcore_avail = 1;
```

阅读手记：

4.3.5 不同目标具备不同 timescale 是否合适？

我们可以通过 \`timescale 宏定义或 timeunit、timeprecision，针对个别模块、接口、类来设定时间的单位和精度，所以，不同的 timescale 也是允许的。我们可以通过系统函数 $printtimescale()来确认当前域的时间单位、时间精度。在打印时间时，也可以从全局的角度设定打印时间的显示方式。在编译时可能有若干目标没有被设定时间单位、时间精度，那么在编译时应该使用-timescale=timeunit/timeprecision 参数（以 VCS 为例）为其设定默认的时间单位、时间精度等参数。

关键词：
timescale，timeunit，timeprecision，$printtimescale()，$timeformat()

避坑指南：
如果某个模块、接口、类对时间单位、时间精度有特定要求，就应该在它的域内设定时间单位和时间精度。如果要在某个任务处延长指定的时间长度，那么应该给定具体的时间（比如避免使用#5，而是使用#5ns）。另外，除非必要，应避免覆盖设计中的时间单位和时间精度，即避免使用-override_timescale=timeunit/timeprecision 参数（以 VCS 为例），因为这可能影响目标在时序上的具体表现。

参考代码：tb_timescale_different_set.sv

```
module rkv_mod1;
  timeunit 1ps;
  timeprecision 1ps;
  initial begin
    $printtimescale();
    #5;
    $display("Current time is %0t", $time);
  end
endmodule

module rkv_mod2;
  timeunit 1ns;
  timeprecision 1ps;
  initial begin
    $printtimescale();
    #5;
    $display("Current time is %0t", $time);
  end
endmodule

module rkv_mod3;
  initial begin
    // timescale is achieved from default or configured while compiling
    $printtimescale();
    #5;
    $display("Current time is %0t", $time);
  end
endmodule
```

```
`timescale 100ps/1ps
module tb;
  rkv_mod1 m1();
  rkv_mod2 m2();
  rkv_mod3 m3();
  initial begin
    // specify time display format
    $timeformat(-12, 2, "ps", 10);
    #5;
    $display("Current time is %0t", $time);
  end
endmodule
```

仿真结果：

```
TimeScale of rkv_mod1 is 1 ps / 1 ps
TimeScale of rkv_mod2 is 1 ns / 1 ps
TimeScale of rkv_mod3 is 10 ps / 1 ps
Current time is 5.00ps
Current time is 50.00ps
Current time is 500.00ps
Current time is 5000.00ps
```

阅读手记：

4.3.6 能否对其他组件执行的 raise objection 强行操作 drop objection？

我们建议在某个 phase 执行过程中，尽可能少地由某些 component 或 object 去执行 raise objection 操作，因为这会增加后期调试的难度，也会对仿真结束的条件做出更多限制。不过，有时候，验证环境中除了在测试序列中执行 raise objection，可能还会在其他地方主执行 raise objection 操作，那么正常的仿真结束就需要执行全部的 drop objection 操作。但有的测试场景需要在执行特定的 drop objection 操作之后，强行退出仿真。那么按照 objection 机制，想要退出仿真的安全方式仍然是执行其他 drop objection 操作。

这里我们给出一段可以复用的、用来强制执行所的 drop objection 的函数：

```
function void force_drop_objections(uvm_phase phase);
  uvm_objection objection = phase.phase_done;
  uvm_object objectors[$];
  objection.get_objectors(objectors);
  foreach(objectors[i]) begin
    phase.drop_objection(objectors[i]);
```

```
      end
    endfunction
```

也可以像参考代码一样,使用 uvm_phase::clear() 来完成当前 phase 的 drop objection 操作。

关键词:

objection,raise,drop

避坑指南:

raise objection 操作和 drop objection 操作都是基于每一个 phase 的。尽可能精简这些 raise objection 操作和 drop objection 操作的数目以简化结束仿真的条件。

参考代码: tb_objection_clear_drop.sv

```
package rkv_pkg;
  import uvm_pkg::*;
  `include "uvm_macros.svh"
  class rkv_component extends uvm_component;
    `uvm_component_utils(rkv_component)
    function new(string name = "rkv_component",
                 uvm_component parent = null);
      super.new(name, parent);
    endfunction
    task run_phase(uvm_phase phase);
      super.run_phase(phase);
      phase.raise_objection(this);
      #5ns;
      `uvm_info("RUN", "phase running", UVM_LOW)
      phase.drop_objection(this);
    endtask
  endclass
  class objection_clear_drop_test extends uvm_test;
    rkv_component comp[2];
    `uvm_component_utils(objection_clear_drop_test)
    function new(string name = "objection_clear_drop_test",
                 uvm_component parent = null);
      super.new(name, parent);
    endfunction
    function void build_phase(uvm_phase phase);
      super.build_phase(phase);
      foreach(comp[i])
        comp[i] = rkv_component::type_id::create (
                    $sformatf("comp[%0d]",i), this);
    endfunction
    task run_phase(uvm_phase phase);
      super.run_phase(phase);
      phase.raise_objection(this);
      `uvm_info("RUN", "phase running", UVM_LOW)
      #3ns;
```

```
                `uvm_info("OBJECTION", $sformatf("RUN phase objection totally
raised count = %0d before phase clear",phase.phase_done.get_objection_total()),
UVM_LOW)
            // objection clear recursively
            phase.clear();
            `uvm_info("OBJECTION", $sformatf("RUN phase objection totally
raised count = %0d after phase clear",phase.phase_done.get_objection_total()),
UVM_LOW)
            phase.drop_objection(this);
        endtask
    endclass
endpackage

module tb;
    import uvm_pkg::*;
    `include "uvm_macros.svh"
    import rkv_pkg::*;
    initial run_test("objection_clear_drop_test");
endmodule
```

仿真结果：

```
    UVM_INFO @ 0: reporter [RNTST] Running test objection_clear_drop_test...
    UVM_INFO tb_objection_clear_drop.sv(30) @ 0: uvm_test_top [RUN] phase
running
    UVM_INFO tb_objection_clear_drop.sv(32) @ 3000: uvm_test_top [OBJECTION]
RUN phase objection totally raised count = 3 before phase clear
    UVM_WARNING @ 3000: run [OBJTN_CLEAR] Object 'common.run' cleared
objection counts for run
    UVM_INFO tb_objection_clear_drop.sv(35) @ 3000: uvm_test_top [OBJECTION]
RUN phase objection totally raised count = 0 after phase clear
```

阅读手记：

4.3.7 如何在回归测试中减少冗余的测试用例？

对于随机种子可能影响激励数据内容，继而对覆盖率产生贡献的测试用例，可以在使用多个种子完成第一次回归测试之后，合并覆盖率并进行测试用例升级（testcase grading）。这样可以分析哪些种子对覆盖率有贡献，哪些种子对覆盖率没有贡献，记录下来这些种子的信息，在下一次测试时指定种子号即可实现仿真资源的有效利用。目前的回归工具（Questa VRM、Synopsys ExecutionManager、Cadence vManager）都具备测试用例升级的功能。

关键词：

regression，coverage，testcase grading

避坑指南：

SystemVerilog 语言本身并没有原生的策略可以在随机变量与功能覆盖率之间形成智能映射，而目前已有的仿真器也难以自动化地实现这一点。不过，已经有一些工具可以协助完成覆盖率的高效收敛，例如，Mentor inFact 和 Cadence PerSpec 均可以实现随机数据的高效生成，更有效地填充功能覆盖率。同时，Cadence Xcelium 的最新版本也已引入智能化，可以基于原有的覆盖率和测试数据来指导产生更有效的激励数据。

阅读手记：

4.3.8 验证环境遇到 reset 时如何协调各验证组件？

首先需要一个监听组件，它能够在监听到 reset 事件后调用 *uvm_phase::jump(PHASE)* 函数，让所有验证结构中的组件立即且统一地跳转至某个 phase（如 reset_phase）。尽管 *uvm_phase::jump()* 函数可以自己实现 drop objection 来服务 phase 跳转，但仍然有一些额外的操作需要验证环境结构层面的考虑，即 phase 跳转前各组件是否需要为响应 reset 做好准备。

如果验证环境在一开始构建时就考虑到了这一点，那么它需要考虑为 sequence、driver、monitor、scoreboard 等组件准备好监听 reset 事件和做出响应的办法。driver 和 monitor 在监听到 reset 时，正常跳转至 reset phase，同时对自身的部分成员做初始化处理；sequence 在监听到 reset 时，由于它不参与 phase jump，它需要在继续、放弃或重新发送激励之间做出选择；scoreboard、reference model 在监听到 reset 时，需要模拟设计做出响应，例如，初始化部分成员变量、清空内部数据缓存等，然后进行 phase jump，RGM 也需要响应 reset 而进行复位处理，以便与设计中的寄存器保持一致。

关键词：

reset，uvm_phase，jump

避坑指南：

如果验证环境一开始没有考虑对 reset 做响应处理而使用了 run_phase（不是 main_phase），那么在后续对 reset 响应的代码完善中，可以将以上处理思路补充到 run_phase 中，也可以将 run_phase 统一修改为 main_phase，再使用 phase jump 的思路来完善代码。

参考代码： tb_phase_jump.sv

```
module tb;
  import uvm_pkg::*;
  `include "uvm_macros.svh"
  class component extends uvm_component;
```

```systemverilog
      bit reset_jump = 0;
   `uvm_component_utils(component)
   function new(string name = "component", uvm_component parent = null);
      super.new(name, parent);
   endfunction
   task reset_phase(uvm_phase phase);
      phase.raise_objection(this);
      `uvm_info("RESET", "phase running", UVM_LOW)
      #1ns `uvm_info("RESET",
                  $sformatf("@%0t phase stepping..", $time), UVM_LOW)
      phase.drop_objection(this);
   endtask
   task configure_phase(uvm_phase phase);
      phase.raise_objection(this);
      `uvm_info("CONFIGURE", "phase running", UVM_LOW)
      #1ns `uvm_info("CONFIGURE",
                  $sformatf("@%0t phase stepping..", $time ), UVM_LOW)
      phase.drop_objection(this);
   endtask
   task main_phase(uvm_phase phase);
      phase.raise_objection(this);
      `uvm_info("MAIN", "phase running", UVM_LOW)
      fork
        repeat(3) begin
          #1ns `uvm_info("MAIN",
                  $sformatf("@%0t phase stepping..", $time), UVM_LOW)
        end
        begin
          if(this.get_name() == "comp[0]" && !reset_jump) begin
            `uvm_info("MAIN", "phase jumpping to reset phase", UVM_LOW)
            reset_jump = 1;
            #1ns phase.jump(uvm_reset_phase::get());
            #1ns;
          end
        end
      join_any
      disable fork;
      phase.drop_objection(this);
   endtask
endclass
class phase_jump_test extends uvm_test;
  `uvm_component_utils(phase_jump_test)
  component comp[2];
  function new(string name = "phase_jump_test",
               uvm_component parent = null);
      super.new(name, parent);
  endfunction
```

```
          function void build_phase(uvm_phase phase);
            foreach(comp[i])
          comp[i] = component::type_id::create(
                    $sformatf("comp[%0d]", i), this);
          endfunction
          task reset_phase(uvm_phase phase);
            phase.raise_objection(this);
            `uvm_info("RESET", "phase running", UVM_LOW)
            phase.drop_objection(this);
          endtask
          task configure_phase(uvm_phase phase);
            phase.raise_objection(this);
            `uvm_info("CONFIGURE", "phase running", UVM_LOW)
            phase.drop_objection(this);
          endtask
          task main_phase(uvm_phase phase);
            phase.raise_objection(this);
            `uvm_info("MAIN", "phase running", UVM_LOW)
            repeat(2) #1ns `uvm_info("MAIN",
                    $sformatf("@%0t phase stepping..", $time), UVM_LOW)
            phase.drop_objection(this);
          endtask
        endclass
        initial run_test("phase_jump_test");
      endmodule
```

仿真结果:

```
    UVM_INFO @ 0: reporter [RNTST] Running test phase_jump_test...
    UVM_INFO tb_phase_jump.sv(12) @ 0: uvm_test_top.comp[0] [RESET] phase
running
    UVM_INFO tb_phase_jump.sv(12) @ 0: uvm_test_top.comp[1] [RESET] phase
running
    UVM_INFO tb_phase_jump.sv(53) @ 0: uvm_test_top [RESET] phase running
    UVM_INFO tb_phase_jump.sv(13) @ 1000: uvm_test_top.comp[0] [RESET] @1000
phase stepping..
    UVM_INFO tb_phase_jump.sv(13) @ 1000: uvm_test_top.comp[1] [RESET] @1000
phase stepping..
    UVM_INFO tb_phase_jump.sv(18) @ 1000: uvm_test_top.comp[0] [CONFIGURE]
phase running
    UVM_INFO tb_phase_jump.sv(18) @ 1000: uvm_test_top.comp[1] [CONFIGURE]
phase running
    UVM_INFO tb_phase_jump.sv(58) @ 1000: uvm_test_top [CONFIGURE] phase
running
    UVM_INFO tb_phase_jump.sv(19) @ 2000: uvm_test_top.comp[0] [CONFIGURE]
@2000 phase stepping..
    UVM_INFO tb_phase_jump.sv(19) @ 2000: uvm_test_top.comp[1] [CONFIGURE]
```

@2000 phase stepping..
        UVM_INFO tb_phase_jump.sv(24) @ 2000: uvm_test_top.comp[0] [MAIN] phase running
        UVM_INFO tb_phase_jump.sv(24) @ 2000: uvm_test_top.comp[1] [MAIN] phase running
        UVM_INFO tb_phase_jump.sv(63) @ 2000: uvm_test_top [MAIN] phase running
        UVM_INFO tb_phase_jump.sv(31) @ 2000: uvm_test_top.comp[0] [MAIN] phase jumping to reset phase
        UVM_INFO tb_phase_jump.sv(64) @ 3000: uvm_test_top [MAIN] @3000 phase stepping..
        UVM_INFO tb_phase_jump.sv(27) @ 3000: uvm_test_top.comp[0] [MAIN] @3000 phase stepping..
        UVM_INFO ${UVM_HOME}/base/uvm_phase.svh(1556) @ 3000: reporter [PH_JUMP] phase main (schedule uvm_sched, domain uvm) is jumping to phase reset
        UVM_WARNING @ 3000: main_objection [OBJTN_CLEAR] Object 'uvm_top' cleared objection counts for main_objection
        UVM_INFO tb_phase_jump.sv(12) @ 3000: uvm_test_top.comp[0] [RESET] phase running
        UVM_INFO tb_phase_jump.sv(12) @ 3000: uvm_test_top.comp[1] [RESET] phase running
        UVM_INFO tb_phase_jump.sv(53) @ 3000: uvm_test_top [RESET] phase running
        UVM_INFO tb_phase_jump.sv(13) @ 4000: uvm_test_top.comp[0] [RESET] @4000 phase stepping..
        UVM_INFO tb_phase_jump.sv(13) @ 4000: uvm_test_top.comp[1] [RESET] @4000 phase stepping..
        UVM_INFO tb_phase_jump.sv(18) @ 4000: uvm_test_top.comp[0] [CONFIGURE] phase running
        UVM_INFO tb_phase_jump.sv(18) @ 4000: uvm_test_top.comp[1] [CONFIGURE] phase running
        UVM_INFO tb_phase_jump.sv(58) @ 4000: uvm_test_top [CONFIGURE] phase running
        UVM_INFO tb_phase_jump.sv(19) @ 5000: uvm_test_top.comp[0] [CONFIGURE] @5000 phase stepping..
        UVM_INFO tb_phase_jump.sv(19) @ 5000: uvm_test_top.comp[1] [CONFIGURE] @5000 phase stepping..
        UVM_INFO tb_phase_jump.sv(24) @ 5000: uvm_test_top.comp[0] [MAIN] phase running
        UVM_INFO tb_phase_jump.sv(24) @ 5000: uvm_test_top.comp[1] [MAIN] phase running
        UVM_INFO tb_phase_jump.sv(63) @ 5000: uvm_test_top [MAIN] phase running
        UVM_INFO tb_phase_jump.sv(64) @ 6000: uvm_test_top [MAIN] @6000 phase stepping..
        UVM_INFO tb_phase_jump.sv(64) @ 7000: uvm_test_top [MAIN] @7000 phase stepping..

阅读手记：

4.3.9 仿真出现错误信息时如何让仿真停止？

对于 UVM 的报错信息，如果是 UVM_ERROR 级别，那么错误信息默认会累积，但是不会暂停（STOP），或不会退出仿真（EXIT）。而对于 UVM_FATAL，仿真会直接退出（EXIT），且无法继续仿真。对于 UVM_ERROR 级别的信息，往往还希望能够在报错数量累积到一定数目时，暂停仿真保留现场，在仿真出错的时间及时进行调试。这时就可以通过 *uvm_root::get().set_report_max_quit_count()* 来设置错误消息累积的最大数目，让仿真退出，也可以通过在仿真时添加参数 +*UVM_MAX_QUIT_COUNT=<count>* 来控制。

关键词：

UVM_ERROR，UVM_FATAL，UVM_MAX_QUIT_COUNT

避坑指南：

有时希望在报告 UVM_FATAL 信息时可以继续执行一段时间而不是直接退出仿真，那就需要修改特定消息严重等级（以及特定消息 ID）的操作。可以通过 *uvm_root::get().set_report_severity_action_hier()* 或 *uvm_root::get().set_report_severity_id_ action_hier()* 这样的函数，先将 UVM_FATAL 级别信息的动作修改为 UVM_COUNT，以便在消息报告出来时不会直接退出仿真。

参考代码：tb_error_count_stop.sv

```
module tb;
  import uvm_pkg::*;
  `include "uvm_macros.svh"
  class error_count_stop_test extends uvm_test;
    `uvm_component_utils(error_count_stop_test)
    function new(string name = "error_count_stop_test",
                 uvm_component parent = null);
      super.new(name, parent);
    endfunction
    function void end_of_elaboration_phase(uvm_phase phase);
      uvm_root::get().set_report_max_quit_count(5);
      uvm_report_server::get_server().set_max_quit_count(4);
      // uvm_root::get().set_report_severity_action_hier(
      //                 UVM_FATAL, UVM_DISPLAY | UVM_COUNT);
      // same effect as statement above 'set_report_severity_action_hier'
      uvm_root::get().set_report_severity_id_action_hier(
                      UVM_FATAL, "FATCNT", UVM_DISPLAY | UVM_COUNT);
```

```
    endfunction
    task run_phase(uvm_phase phase);
      phase.raise_objection(this);
      fork
        for(int i=1; i<=10; i++) begin
          #1ns `uvm_error("ERRCNT",
                    $sformatf("Error report count number is %0d", i))
        end
        for(int i=1; i<=10; i++) begin
          #1ns `uvm_fatal("FATCNT",
                    $sformatf("Fatal report count number is %0d", i))
        end
      join
      phase.drop_objection(this);
    endtask
  endclass
  initial run_test("error_count_stop_test");
endmodule
```

仿真结果：

```
    UVM_INFO @ 0: reporter [RNTST] Running test error_count_stop_test...
    UVM_ERROR tb_error_count_stop.sv(20) @ 1000: uvm_test_top [ERRCNT] Error
report count number is 1
    UVM_FATAL tb_error_count_stop.sv(23) @ 1000: uvm_test_top [FATCNT] Fatal
report count number is 1
    UVM_ERROR tb_error_count_stop.sv(20) @ 2000: uvm_test_top [ERRCNT] Error
report count number is 2
    UVM_FATAL tb_error_count_stop.sv(23) @ 2000: uvm_test_top [FATCNT] Fatal
report count number is 2
```

阅读手记：

4.3.10　Verilog 如何实现在相同结构中采用不同设计模块？

Verilog 本身能够通过在不同 configuration 中就同一个设计层次中的模块或实例，选择不同的设计模块。要这么做，需要将不同的设计模块（同名模块、不同实现内容）预先编译到不同的库中，然后在编译 configuration 时在其内部指定从哪个库获得目标设计模块，继而完成整体的 elaboration 链接过程。

关键词：
configuration，elaboration，library

避坑指南：

这样的处理方式，是 Verilog、VHDL 这些语言本身的功能，但需要有不同的 configuration、library 以及不同的链接目标（仿真可执行对象）。在仿真时，同样要指定不同的仿真可执行对象，才能够采用不同的设计模块。

参考代码： tb_verilog_config（文件夹）

```systemverilog
// m2.sv
module m2; // m2 type-1 compiled into work library
  logic v1, v2;
  initial begin
    $display("This is m2");
  end
endmodule

// m2_dummy.sv
module m2; // m2 type-2 compiled into tblib library
  initial begin
    $display("This is dummy m2");
  end
endmodule

// m1.sv
module m1;
  m2 m2_inst();
endmodule

// top.sv
module top;
  m1 m1_inst();
endmodule

// config.sv
config cfg1; // take m2 type-1
  design work.top;
  default liblist work;
endconfig

config cfg2; // take m2 type-2
  design work.top;
  default liblist work;
  instance top.m1_inst.m2_inst liblist tblib;
endconfig
```

阅读手记：

4.3.11 仿真器如何实现在相同的结构中采用不同的设计模块？

这里以 VCS 为例。VCS 提供一种仿真过程中的动态设计重配置特性（dynamic reconfiguration at runtime）。相比于 Verilog、VHDL 语言特性本身支持的 configuration 而言，这一特性通过恰当的环境准备和编译，在仿真运行时，仅通过命令项即可达到替换目标实例的效果。这种方式便于去挖空、替换设计结构中的某些组成部分。除了同样需要准备一个可替代的设计模块（substitution module），还需要在 elaobration 阶段提供动态配置命令文件（dynelab.config）。在这种模式下，只会生成一个仿真目标可执行文件。在接下来的仿真阶段，可以选择使用替代模块（添加运行时命令 *-dynaconfig dynarun.config*），也可以不添加动态配置命令（即使用原有设计结构）。

关键词：
VCS，dynamic，reconfiguration，runtime

避坑指南：
如果在验证环境中实现该种动态配置，那么需要考虑在不同设计配置情况下，环境和测试用例的适配性。例如，如果设计实例被挖空替换，那么应该避免再去索引任何原实例中的信号、变量或其他内容。

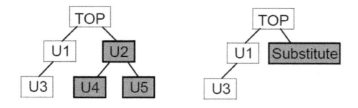

参考代码： V3 验证课程代码

```
// dynaelab.config
replace module { CortexM3Integration } with
        module { CortexM3Integration_sbst };

// dynarun.config
mcss_top_tb.dut.u_cortexm3

// Makefile
ifeq ($(DYNCFG), 1)
COMP_OPTS  += +define+DYNCFG_M3
ELAB_OPTS  += +optconfigfile+dynaelab.config
RUN_OPTS   += -dynaconfig dynarun.config
endif
```

阅读手记：